T0221442

Praise for *Understanding the Metaverse*

"The Metaverse is one of those generational transformative technologies that will disrupt all aspects of our lives. Nick's book is a thorough guide that makes it understandable, and more importantly, actionable for a general business audience."
                    —**Alvin Wang Graylin**, China President, HTC

"The metaverse is the future of fun, work, learning, and more and Nick Rosa's book is an excellent guide leading you to that future."

—**Paul J. Zak**, author of *Immersion: The Science of the Extraordinary and the Source of Happiness*

UNDERSTANDING THE METAVERSE

UNDERSTANDING THE METAVERSE

A BUSINESS AND ETHICAL GUIDE

NICK ROSA

WILEY

This work was produced in collaboration with Write Business Results Limited. For more information on Write Business Results' business book, blog, and podcast services, please visit their website: www.writebusinessresults.com, email us on info@ writebusinessresults.com or call us on 020 3752 7057.

Registered office
John Wiley & Sons, Inc., 111 River Street, Hoboken, NJ 07030, USA

John Wiley & Sons Ltd, The Atrium, Southern Gate, Chichester, West Sussex, PO19 8SQ, United Kingdom

Editorial Office
John Wiley & Sons Ltd, The Atrium, Southern Gate, Chichester, West Sussex, PO19 8SQ, United Kingdom

For details of our global editorial offices, customer services, and more information about Wiley products visit us at www.wiley.com.

Wiley also publishes its books in a variety of electronic formats and by print-on-demand. Some content that appears in standard print versions of this book may not be available in other formats.

Designations used by companies to distinguish their products are often claimed as trademarks. All brand names and product names used in this book are trade names, service marks, trademarks or registered trademarks of their respective owners. The publisher is not associated with any product or vendor mentioned in this book.

Limit of Liability/Disclaimer of Warranty
While the publisher and authors have used their best efforts in preparing this work, they make no representations or warranties with respect to the accuracy or completeness of the contents of this work and specifically disclaim all warranties, including without limitation any implied warranties of merchantability or fitness for a particular purpose. No warranty may be created or extended by sales representatives, written sales materials or promotional statements for this work. The fact that an organization, website, or product is referred to in this work as a citation and/or potential source of further information does not mean that the publisher and authors endorse the information or services the organization, website, or product may provide or recommendations it may make. This work is sold with the understanding that the publisher is not engaged in rendering professional services. The advice and strategies contained herein may not be suitable for your situation. You should consult with a specialist where appropriate. Further, readers should be aware that websites listed in this work may have changed or disappeared between when this work was written and when it is read. Neither the publisher nor authors shall be liable for any loss of profit or any other commercial damages, including but not limited to special, incidental, consequential, or other damages.

Library of Congress Cataloging-in-Publication Data

Names: Rosa, N. (Nicola) author.
Title: Understanding the metaverse : a business and ethical guide
  / Nicola Rosa.
Description: First Edition. | Hoboken, NJ : Wiley, 2023. | Includes index.
Identifiers: LCCN 2022032492 (print) | LCCN 2022032493 (ebook) | ISBN
  9781119911807 (cloth) | ISBN 9781119913641 (adobe pdf) | ISBN
  9781119913634 (epub)
Subjects: LCSH: Shared virtual environments—Social aspects. | Virtual
  reality. | Business—Computer programs. | Business ethics.
Classification: LCC HM742 .R667 2023 (print) | LCC HM742 (ebook) | DDC
  302.30285—dc23/eng/20220804
LC record available at https://lccn.loc.gov/2022032492
LC ebook record available at https://lccn.loc.gov/2022032493

Cover Design: Wiley
Cover Image: © solarseven/Getty Images
Printed and bound by CPI Group (UK) Ltd, Croydon, CR0 4YY

C9781119911807_220922

CONTENTS

ACKNOWLEDGMENTS AND DEDICATION

I want to dedicate this book first and foremost to my parents Lina and Tommaso, who always enabled and encouraged me to pursue my passions and interests in full; and to my dear brother Francesco, whose determination, resilience and moral compass are a constant source of inspiration to me.

To Professor Tom Furness, for bestowing on me the great honour of writing the foreword of my book, and whose greatness, kindness and incredible career keep being a constant source of inspiration for thousands of leaders in the Metaverse business – many of whom have been his students at the University of Washington.

To the incredible pool of industry experts who contributed to some of the chapters of this book – Dr. Fadi Chehimi, Maria Mazzone, Jamie Solomon and my fellow geek Bob Gerard. You all fill my life with knowledge, inspiration and fun, and I feel blessed in having the privilege of calling you friends. I'll always be grateful for your help on this book.

To all the people who, during my career, gave me a platform to grow professionally and organically without forcing my path in any way, and taught me how to find my own steps. To the mentors and managers who became dear friends and an integral part of my life, Rocco 'Ringo' Anaclerio, Alex Catlin, Ian Hunt, Cian O'Hare, Marc Carrel-Billiard and Dan Guenther.

To my closest, dearest friends Giuseppe, Cristina, Ben, Verena, Giada, Gianluca, Georgia, Max, Erik, Pasco, Fabio, Antonella, Daniel, Sam, Maria and Nikos, who are my extended family, whom I love deeply and I could never live without.

To my editor at Wiley, Annie Knight, who gave me the opportunity to write this book; and to Georgia Kirke at Write Business Results and her teammates Ivan Meakins and Katherine Lewis, who helped with the process.

And, last but not least, I want to thank you for picking up my book. I hope you'll find it interesting, informative, and, why not, fun! As in every Metaverse adventure, in the end, the thing that will mostly remain in your memory about this book will be the experience of reading it, and I hope you'll have a good one.

FOREWORD

"Tom Furness is an amalgam of Professor, Inventor, Virtual Reality Pioneer and Entrepreneur in a professional career that spans 56 years. He is currently a professor of Industrial and Systems Engineering at the University of Washington (UW), Seattle, Washington, USA. Dr. Furness has published over 400 papers and conference proceedings and started 27 companies, two of which are traded on NASDAQ at a market capitalization of > $ 12 B (USD). He is a Fellow in the IEEE and a member of the Computer Society and Photonics Society of the IEEE. He is the founder and chairman of the Virtual World Society, a non-profit for extending virtual reality as a learning system for families and other humanitarian applications. He has received four lifetime achievement awards for his work in photonics, electro-optics, human interface technology, and education and is considered the 'grandfather' of virtual reality and augmented reality."

Oh no . . . not another book about the Metaverse! That was my first reaction when Nick Rosa approached me to contribute a foreword to his book. Of course, I was honoured that he would ask me, but REALLY. I envisioned another tome saying, 'Golly, gee, wow, now look at what we can achieve by melding the Internet with virtual reality, artificial intelligence, cryptocurrency, non-fungible tokens, etc. – just think of the applications!'

However, the more I thought about this and actually read a draft of Nick's book, the more I realised that it IS a big deal! As we look back at the evolution of media, each advance from the printing press to telegraph, radio, film, television and the Internet was transformational. They didn't make the old way of doing things obsolete (we still read books, listen to the radio, watch TV and movies) as much as they added to our intellectual circumference by unfolding new dimensions that expand our experiences and connectivity. As radio was able to transport our ears to another place, television our eyes and the Internet our brains (or lack of brains), the Metaverse has become the ultimate transportation systems for our minds. As such, and with its global reach, it is key to uniting us to solve the problems that our civilisation faces and save the earth upon which we live.

From his experience as a thought leader in digital transformation at multiple global consulting services, Nick is uniquely positioned to see the potential impact of this emergence, especially in business, enterprise, education and government. In this book, Nick's intent was not to produce a technical how-to-do book as much as to provide a primer for business and government leaders to get a taste of what is coming and to prepare for the new emergence.

I keep thinking about the recent deployment of the Webb Space Telescope. The purpose of this new telescope is to gather and focus light . . . a special kind of light that we don't see with our eyes. Furthermore, it is important where the telescope is aimed with its many mirrors, so that its vision is not impaired or contaminated by the background, but instead is able to look outward and see afresh. To extend this as a meta-phor, Nick's book is like the Webb Telescope. He has written it to focus our minds to help us see beyond the convention of the Internet and other historical media. In this way, its purpose is to enlighten us, so that we can grow a new comprehension of what lies ahead and plan for it.

These days, it is easy to become intoxicated (or should I say seduced?) by technology. We believe it can do anything, but believing this is so doesn't necessarily make us happy. When we contemplate the true power of the Metaverse to move minds, we have to give pause and realise that we are playing with fire, as the impact on the brain is unprecedented. It's true that such a fire can illuminate, keep us warm and generate steam to lift humanity, but, if not used carefully, it can also destroy us, this time from the inside out. For these reasons, Nick is quick to add caution to our intoxication because, like fire that can harm or heal, the impact of the Metaverse, good or bad, all depends on the responsible use of this technology.

Although it has been under development for many years, the key enabling technology for the Metaverse is virtual reality (VR). Yet, we still don't know the longitudinal impact of using virtual interfaces over long periods of time. Nor do we understand the power that misinformation conveyed through such an immersive medium can have on discourse, as it is difficult to unlearn what has been experienced in VR. So, this begs the question about the ethics of policing and/or censoring the open exchange of information in the Metaverse. One could envision an artificial intelligence police force that watches every comment, exchange and image shared in that future Metaverse. Do we really want that? These and other ethical questions need to be brought to the forefront as we develop and embrace this new reality. Nevertheless, haven't we experienced this type of conundrum before, like when we split the atom (or fused two atoms into one)? We understood with this advance that we were releasing enormous energy that could fuel our civilisation – or destroy it.

Read this book with these considerations in mind. Thank you, Nick, for this introduction to the new and burgeoning world of the Metaverse.

Tom Furness
Seattle
June 2022

INTRODUCTION

Are you ready to enter the Metaverse? The fact that you're reading this book tells me that you are, at the very least, curious about all it has to offer. Maybe you've simply heard the term thrown around and want to learn what it really is (you're not alone!) and how it could be of use to you and your business. You may be feeling overwhelmed by all the information that's out there and want to cut through the noise, to break this vast concept down into manageable chunks of information that you can digest.

I can tell you that there are a multitude of exciting use cases within the Metaverse already, and that many more will emerge in the coming months and years as we learn how to harness this new technology to enhance our lives and build businesses.

Through this book, you and I are about to embark on a journey of discovery through emerging technology. We are not seeking to reach a specific destination and move quickly from A to B; ours is a journey of exploration, one where there are exciting new things to uncover, as well as threats to tackle, in every corner of our virtual map.

What guides us on this journey is not where we are going – because, truthfully, none of us know where that will be – but your business goals and the values you want to convey to your customers as we take our first baby steps into this new, exciting world.

The Metaverse is like a greyed-out map in an open-world game, such as when you start playing the Nintendo videogame *The Legend of Zelda: Breath of the Wild* – the map is formed, but you don't know what you're going to discover as you start exploring. We are taking our first steps into the Metaverse, and we are not yet sure what characters, obstacles, challenges and opportunities we will find there. We might strike gold or uncover hidden treasures. We also might run into monsters and threats that need to be faced and tackled before we can continue our journey. We could face unexpected obstacles, like mountains that have to be scaled to allow us to see new horizons before us. We may find new communities who welcome and support us, and we will likely encounter trolls that need to be slain.

In the not-too-distant future, businesses across all industries will need to enter the Metaverse, whether they want to or not, much like businesses nowadays need a website and online presence to thrive. What I want to do through this book is to equip you with the tools, knowledge and awareness that you and your business will need as you make your first forays into this brave new digital world.

See this book as a box of loot that you pick up on your digital adventures. Inside, you will find tools, tips and useful strategies to better your understanding of the Metaverse and begin exploring the map.

Together, we will explore what the Metaverse is, how it works, its many use cases that we are already aware of, as well as look at why there are so many more we cannot yet see. I will also introduce you to some of the main threats and challenges that businesses and individuals face as the Metaverse evolves and grows, sharing potential solutions to make this an inclusive virtual world that all of us can enjoy and use to its full potential.

I will also explain some of the key elements of technology that underpin the Metaverse in its current form and which will provide the necessary foundations for its growth and evolution in the future – such as blockchain, cryptocurrency, AI and deep learning. I will build out the picture with you, helping you to turn some of the greyed-out map into a world filled with technicolour and exciting opportunities. By the end of this book, I hope you will feel confident to set out and explore more of this uncharted territory on your own.

Before we land in the Metaverse, however, we're going to look at the role that technology has played in human evolution over the centuries. This is fundamental to understanding why the Metaverse is far from a fad and is, in fact, set to be truly transformational for human society.

CHAPTER 1
TECHNOLOGY AND HUMAN EVOLUTION

Imagine how incredible it must have been the first time people created fire. The awe at its properties; the pain of discovering it burns; the fear of the destruction it can wreak; the excitement over how it can be used to cook, heat and illuminate. Fire, like the invention of the wheel, is just one example of how technology and new discoveries have shaped our society. Throughout the course of human history, technology has played a major role in our evolution as a species.

However, since the first Industrial Revolution, there have been a series of 'technology waves' that have had a significant impact on the evolution of human society. Rather than incremental gains, these have appeared like tsunamis, flooding the world and changing it suddenly and irrevocably.

If we take the invention of the railway, which occurred during the second wave, we can see how this invention rippled out through society. Not only did it affect commerce, industry and tourism, but it also impacted, for instance, the urbanisation of rural areas in the United States. Rail transport made it easier for people to travel into cities and therefore changed the opportunities available to those living in more remote communities.

In more modern times, we have seen a similar shift in our behaviour as a society with the widespread adoption of mobile devices. We use them day in, day out; they have deeply impacted the way we communicate with one another, the way we work and the way we build relationships. How many of you would feel lost and naked without your smartphone in your pocket (I know I do!)?

These waves of technological innovation are not only about the discovery of a new technology. For true transformation to happen, technology, user experience and business models must all converge to create new, truly transformative products and services. This is a natural phenomenon that occurs with any new technology after a given amount of time; it is called 'emergence'.

This is when we see disruptive technology changing society and our way of life. It's not enough to simply innovate with new technology; without the other two components, a well-chiselled user experience and a scalable business model, it won't be adopted at scale and therefore won't have a society-wide impact.

Emergence is what happens when one or multiple use cases using an innovative technology become highly successful and trump all the others that exist at the time.

Disruptive products and services, therefore, have a much broader impact than the technology that enables them. They reverberate around the world on a macro level. Look at the adoption of mobile phones as an example. It started slowly at the beginning of the 1980s, but it's only with the advent of smartphones that we have seen explosive adoption of the technology – thanks to the deep integration of smartphones with ecosystems of services and business models that work (app stores); a great and simple user experience (touchscreens); and the arrival of fast mobile data network technologies (EDGE and 3G) that supported rich media experiences. There are now twice as many smartphones in existence in the world as there are people on the planet[1]. When technology is distributed at this scale, it leads to macroscopic changes in the way society, the economy, welfare, politics and much more works. The next tech revolution we're entering, with the Metaverse and immersive technology, will have a profound impact on the way we work and the way we interact with other people. There will undoubtedly be currently unforeseeable changes to the way we live once this technology is integrated into our lives.

What is the Metaverse?

The easiest way to think about the Metaverse is as the next evolution of the Internet. There will be platforms within the Metaverse that offer content, experiences and the chance to live in different, virtual worlds. These platforms will, eventually, become interconnected. The vision is for us, as users, to be able to move seamlessly from one platform to the next.

The Metaverse will also incorporate augmented reality (AR) content that will be overlaid onto the physical world. The rise of wearable tech devices (which I'll discuss later) will mean that this becomes as natural to us as using Google Maps to navigate our way around a new city. In Chapter 3, we'll explore what the Metaverse will look like, where the concept has evolved from and the different ways its evolution could go.

In order for all of this to be possible, certain technological advances will be required. Many of these are already on their way, but most are being created at the time this book is being written, as you'll learn in this chapter and throughout the rest of this book. Let's start by looking at how technology has already changed our reality, and how it has the power to do so again.

Reality is constantly evolving

The reality we're living in right now is very different to the reality we were living in 15 years ago. In the 2020s, our reality has been shaped by the advent of new technologies, which makes it more complex than our reality from 15 or 20 years ago. In turn, the reality that we're moving towards will be different and more complex than the one we're currently living in.

Right now, if you leave your house without your mobile phone, I expect that you will return home to get it, because your world (your reality) is less meaningful and less experiential without it. In the future, our reality will be enhanced by a layer of digital content and information that is superimposed over our vision.

In our future world, we will be able to customise our reality to whatever we want it to be. If we wake up in the morning and the sky is grey, we will have the ability to transform the sky to blue. We are about to enter a world where we literally become omnipotent in terms of how we're seeing what is around us, which is something we've never had before.

Of course, we've created worlds in videogames, and we've already created fictional realities in movies and in the theatre. Since the dawn of the human race, people have created fictional worlds to tell stories and, soon, for the first time in human history, we will be able to shape reality as we want to see it. This is something that will become more and more prominent in the future. At the time of writing, we can only achieve this through glasses and headphones to allow us to see and hear this alternative reality. But, in the future, maybe we will also be able to use haptics (transmitting and understanding information via touch) or even brain implants to broadcast the information directly in our neural system.

Of all the senses, however, it is sight that is likely to be the most important, certainly initially, because the optic nerve has the largest bandwidth directly to the brain, and therefore visual information is incredibly important to us in terms of decoding the world to our reality.

So much of the content being created for the Metaverse is visual in nature. We're literally designing new 3D worlds, and how they look is just as important as how they function. As the Metaverse is going to be such a visual medium, we have to consider the ways in which it can shape how we perceive the world. A focus on visual content will improve the sense of immersion of the user, blurring the lines and blending the reality of the Metaverse with our physical reality. The more technology evolves to bring our other senses with us into the Metaverse, the more immersive the experiences will become and the more challenging it will become to separate physical from digital reality.

Navigating the ethical maze

The Metaverse has the potential to be truly amazing. We could help kids from anywhere in the world get university education, for instance. We will be able to interact with our loved ones around the world in much more meaningful ways than just using a video call. At the same time, however, the technology could be used for nefarious purposes, such as mass manipulation, or to enable someone to simulate child pornography. On the one hand, you could argue that's better, because no real children will be involved, but on the other it's still fundamentally wrong. Or those who take drugs to escape their current reality could turn to the Metaverse instead, which could become like a drug itself. There are a great number of philosophical and ethical conversations we need to have within society when we are building new realities. We will need to navigate our way through these ethical mazes.

We also have to ask whether these realities are going to have the same value as the real world or not. How are we going to treat them? What's the level of reach of our current legislation in terms of human norms in those realities?

The reason I'm raising these questions and challenges now is that, when we're exploring the business ethics of operating in the Metaverse, we need to be aware not only of the incredible possibilities it presents, but also the potential dangers, so that we can navigate towards this new technology and new era of human reality and evolution with a conscious mind.

Balancing the positive with the negative

There is always a backlash to the arrival of new technology, and there are also challenges that come with the introduction and adoption of technology on a societal level. Take the

Internet and social media, for example, where there have been various concerns and issues surrounding security, privacy and data, such as the multiple hackings and cyber warfare attacks that happened in the last decade or the Cambridge Analytica scandal that enabled mass manipulation and attempts to influence sovereign state elections from an enemy state. On the other hand, the Internet has enabled us to work remotely, access information from wherever we are in the world and given more people access to education.

Whenever a new technology is on the cusp of development and adoption, it's important to study what the potential backlash to that technology could be and to identify what dangers this technology could pose to society in the future. In doing so, we can also create a common view of what's right and wrong when it comes to the use of this technology, and potentially involve governments at this stage as well (which I'll cover in greater detail later in this book).

The Metaverse is the next wave building that has the power to change the course of society. According to Dorothy Neufeld[2], there have been six waves of innovation from the eighteenth century through to 2020, starting with the likes of water power (hydro-power dams), ability to improve the production of textiles and iron, up to the fifth wave of digital networks, software and new media. At the time of writing, we are entering a sixth wave of innovation, encompassing clean tech, the Internet of Things (IoT) and artificial intelligence (AI), and robots and drones. I believe this sixth wave will also include the beginning of a new way for people to interact with each other, using virtual three-dimensional worlds, and make it possible to conduct business through virtual products and services – which will also bleed into our reality, connecting the digital and the real in a seamless experience.

During each wave of innovation, old technology is disrupted as new technology takes hold. This has an impact on everything from the distribution of people within different countries to the destruction of some jobs and the creation of others. The Metaverse is coming, and we need to be prepared for the opportunities and disruptions it brings.

True value drives technological adoption

When we look back through history, we can see that no technology achieves widespread adoption without being truly transformational. If a new technology doesn't deliver real value for users, it is just a fad and will eventually disappear. Please note that this can happen because of the intrinsic value of the technology or simply because of its lack of maturity for mass-market adoption. The use cases for each piece of technology must be solid and should create measurable value for consumers as well as the enterprise producing or selling it.

Railways were transformational because they enabled goods and people to be transported across vast distances, at scale, very efficiently. Mobile phones added transformational value by enabling us to contact anyone, anywhere in the world. Applications on smartphones almost gave us superpowers – just think about how incredible it is to be able to hail a taxi at the press of a button!

The true transformational power and mass-market diffusion of technologies like immersive tech and virtual worlds will happen when truly transformational use cases start to appear. For example, I might be able to discern your emotions during a conversation I'm having with you because my AR glasses are picking up your micro-expressions to tell me whether you're happy or unhappy, and an AI assistant is then able to suggest the next best action I can take to steer the conversation in a particular way.

This might sound like something straight out of sci-fi, but this could be possible in a matter of years. Of course, technological developments like this are fraught with ethical complications. These are the monsters we need to confront as we explore the Metaverse. We cannot simply change course to avoid them, because they will appear in every direction. The ethics surrounding how we use this new technology is an essential consideration for any business seeking to enter the Metaverse, and this is a topic I will cover in greater detail as we move through this book.

New technology seeks to make the impossible possible. To do this, we need to uncover the unique capabilities linked to a specific device and identify how we can use these for maximum effect. In doing so, it is possible to create something completely new that was impossible to achieve before. One of the challenges for those who are creating transformational use cases using new technologies is what's known as *skeuomorphism*, or the imitation of the user experience (look and feel) of what we are already familiar with in our everyday life.

It takes a certain kind of person to be able to see beyond the obvious, which is why, whenever a new technology is developed, there is always a degree of imitation of previous technologies and mediums – I'll highlight some examples in the next chapter. Usually, the truly transformational use cases for new technologies only appear five to ten years after their invention, when we have a much better understanding of the user experience, capability and what's really possible.

From vision to reality

At the beginning of this process, we are literally just scratching the surface and only just starting to understand the art of the possible. As we explore different use cases, one will usually peak, and this is what drives a technology into truly transformative territory.

Virtual reality (VR) is a good example. Although there is a lot of buzz around this at present, it's far from a new concept. In fact, in the 1980s and early 1990s, everyone thought VR was going to conquer the world. We had movies like *Tron*, where someone could inhabit a virtual world (albeit they couldn't leave, and the games take a deadly turn!), and *Total Recall* (who could forget Arnie's visit to Mars?!), where memories and experiences could be implanted in our brains. There was a real frenzy surrounding VR, but then the Internet arrived and stole the show.

The fact was that the technology available wasn't anywhere near sophisticated enough to deliver an immersive VR experience like the movies portrayed. For VR to come into its own, we need multiple complex technological advancements to allow users to see an immersive and realistic virtual world along with a comfortable form factor for the wearable devices.

The problem that VR had in its early days was that technology wasn't up to scratch. In fact, VR technology has been lurking in our lives for about 50 years, but it's only in the last five years or so that the technology has achieved a level of sophistication that enabled its mass-market productisation and is kickstarting a process that could lead to mass-market adoption. Even so, we're not there yet, because not everybody has VR headsets, and we are certainly a long way from having memories of visiting Mars implanted into our brains!

Still, this illustrates how long it can take from a technology's initial development to its widespread use and therefore its ability to transform society. The realisation of the vision that the Metaverse is, of course, about much more than achieving the mass-market adoption of AR and VR. The final goal to realise the commonly agreed vision of the Metaverse is having an **interconnected network of 3D worlds** and platforms accessible by any device (including computers, TV, mobile phones, AR/VR headsets, etc.). This network should support the cross-platform exchange of digital goods and currencies; be governed by a constellation of commonly agreed and widely adopted standards, regulations, and identity systems; and be powered by high-speed, low-latency networking infrastructure and by highly efficient computing clusters to allow thousands, if not millions, of concurrent users to interact with each other. A culmination of multiple things – and all of this need to happen at the same time in order for it to take off.

All of these small waves are gradually converging to swell into a tsunami that will change the face of the world. At the moment, we have thousands of companies, institutions, and associations that are all working on the necessary components to achieve this vision. It is not an easy task, as many of these components are deeply interconnected with multiple other points of the infrastructure.

In the future, for example, the Metaverse will need to be completely rendered or partially processed in the cloud to offload the computational power required from local devices. The challenge of relying solely on a local device is that the processor it contains generates heat that needs to be dissipated in some way. For example, if you have to wear a pair of glasses to access the Metaverse, they have to be thin and socially acceptable, otherwise people won't wear them; and they can't be required to process too much information, otherwise their battery will drain too quickly or, in the worst case scenario, its components will overheat to uncomfortable levels.

This means that processing power needs to be transferred to the cloud – and, in order to communicate with the cloud, you need good network connectivity. Therefore, the network is an incredibly important part of the picture.

The glasses needed to access the Metaverse must not only be comfortable and light, but they also need to be able to render information with high contrast and clear definition even in broad daylight. At present, even looking at the phone screen is difficult when out in the sun. All of this technology needs to be developed in order for the dream of the Metaverse to materialise.

At the time of this writing, we're almost there, but not quite; there are still a few pieces of the puzzle that are missing.

What we are assisting at the moment of writing this book it's a 'moonshot', not dissimilar to what the U.S. President J.F. Kennedy announced during his historical speech – He set the post high, and clear stating that he wanted to get an American citizen to step on the moon within in a decade. In the same way many companies currently operating in the Metaverse business (including Meta), have a clear vision of what they want to achieve and believe that within the next ten years, creating a full version of the Metaverse will be possible from a technical and design standpoint – although with some compromises and surely some unexpected differences from our current vision.

The Metaverse and the wave of sustainable tech

Sustainability also plays a key role in the sixth wave of innovation, which includes the development of clean energy and electric cars, etc. The Metaverse may also play a role in the sustainability topic. This is because Metaverse technologies will enable new ways of meeting people for both work and pleasure, and this will not require the amount of energy that travelling does. Potentially, it will be possible to attend any event, anywhere in the world, without ever having to leave your home.

Physical travel will, of course, still be something we do because we want to explore and experience the world – we are human after all – but being able to travel virtually will make the way we live our lives more sustainable as we will spend less energy. We've already seen how having a telepresence can be beneficial with the rise of remote working; we have become increasingly independent of our geographical locations.

In 2022, technology doesn't yet allow us to feel as though we are really in another geographical location. However, the related technology is going to keep improving and, in the future, what you'll see in Metaverse platforms will be indistinguishable from real life. If you don't believe me, just look at how graphics in video games have evolved in the last 20–30 years. This evolution is inevitable, and we will get there, as you'll discover later in this book.

The impossible will become possible

As technology and its standards develop, what seems impossible to us now will become possible in the future. Just look at the functionality we have on our mobile phones – 20 years ago, this would have been unthinkable. Devices that we can't even think of now (or that have only been present in sci-fi TV shows) will emerge and transform the way we live. Immersive technology has the power to change our lives in ways we can't even imagine.

The Metaverse will use immersive technology and encompass virtual worlds, virtual commerce and non-fungible token (NFTs), but I believe we also need to view the Metaverse from a more holistic perspective. The Metaverse is much more than just virtual worlds, virtual commerce and NFTs; it will bring together everything that is connected to humans using 3D images and immersive technologies.

It will not only allow us to have more immediate and intimate access to information and news, but also to travel and communicate differently. It could give us the ability to record and relive our memories, or even to share those memories with others. This, in turn, could make the world more empathetic because of its immersive nature.

Think about this from the perspective of the emotional impact that watching footage of an event on TV has as compared to reading a magazine article about the event, and then imagine how much more impact would be made if you were able to see that footage from the perspective of someone who has actually lived through the event. While my hope is that this technology will create more empathy in the world, there is also always the danger that it could be weaponised and used to manipulate. These are just some of the ethical issues we will tackle as we move through this book.

With great power comes great responsibility

We can already see how our data is being used to manipulate us via social media platforms. How many tailored ads do you see as you're browsing the Internet every day? How often do you mention a product, experience or topic in passing only to see a related advert pop up on social media the very next day? Large companies are already farming our data (and we give a lot away for free) and selling it to the highest bidder. Everything we do online is already being tracked as far as possible with existing technology, so it is safe to say this will continue to happen if we don't introduce some form of regulation around it.

This is where the danger comes in. The kinds of devices we will be using in the future will obviously provide a great deal of behavioural and biometric data about us, and when this is tied to machine learning algorithms, there is the potential to create highly tailored virtual or digital sales assistants. These persuasive virtual avatars could not only be used to try to sell you products, but could also be repurposed to even influence your political views, for example. Because of the amount of data available about how you behave, talk and even think, it will be very easy to create an incredibly detailed profile of what makes you tick, and therefore to create incredibly persuasive virtual assistants.

As with any new technological innovation, the greater the power it offers, the greater our responsibility to use it sensibly. This is why it's important to be aware of the risks associated with the Metaverse, even as we get excited about the amazing opportunities it presents.

This is especially pertinent in relation to the Metaverse because it will encompass both our reality and our perception of reality. Reality is shaped in our brains and has meaning for us based on who we are and how we perceive it. We also project our own reality outside of ourselves. But what if the reality that is around us can be redesigned to influence our perception of things and manipulate the way that we think? This could obviously go either way in terms of being positive or negative (we don't want to find ourselves in *The Matrix*!).

There are many studies showing that reality is very subjective and that our state of mind can affect the way we perceive the world. However, while the world that we perceive around us is not really physically modifiable based on what we want, it is shaped by how we think.

With these new forms of reality, it will become possible for other people to shape the way in which we see the world, much more than it has been at any time in the past. The flip side to this is that we could enter a world that we design ourselves. For the first time ever, this would allow us to live in the world that we want. At the same time, this technology can also be used to shape the way we think, because our experience creates memory, which in turn creates a different way of thinking. This could have amazing applications for those with mental health challenges, for example.

However, this could also be a very powerful weapon for manipulation, because, if this technology has the capacity to change the way we think, it also has the ability to modify how we perceive reality in the real world. It's an important ethical debate to have, and businesses seeking to operate in the Metaverse – and also institutions and regulatory organisations – need to take responsibility for leading that debate to make the future of the Metaverse sustainable, safe and inclusive for all its users and for our society.

The Metaverse is coming

This book is not here to debate whether the concept of the Metaverse is right or wrong. The Metaverse is inevitable. It is coming, whether we want it to or not, much like the Internet and social media entered our lives and society.

Naturally, there are people saying they don't want to be a part of it, but closing your eyes and burying your head in the sand is not going to help anyone. It is far better if we brace ourselves and prepare for this inevitable new reality. In doing so, we can do our best to create a blueprint for this technology and educate everyone about how to make responsible use of this new technology.

Our evolution shifted gears a long time ago with the advent of the first Industrial Revolution, if not sooner. Since then, our evolution and ability to improve our lives have been connected to our use of technology, whether in terms of agriculture, communication, medicine or even new breakthroughs in biology. Humanity will always move towards the next technological step – it's inevitable (as long as this very process will not cause humanity's extinction).

It is also far better to be aware of what's coming in this respect, so that we can prepare for it. The final vision of what the Metaverse will be is a combination of multiple technologies, standards and regulations coming together. It encompasses AI, 5G (or even 6G), 3D engines, VR and AR hardware, blockchain, networking, cloud and much more. In addition to all of these, there is the adoption of different use cases, for both enterprises and consumers. It's this emergence of factors that leads to transformational technology developments, as I said at the beginning of this chapter.

We've seen this happen multiple times in recent years. Skype emerged when people switched from dial-up Internet connections to broadband and fibre optic connections. Spotify exploded as a mobile app when 3G arrived and the iPhone was launched. Of course, this facilitated not only Spotify, but also music streaming more broadly. Now, with the arrival of 5G and soon also 6G networks, we will have the ability for web AR platforms to thrive and bring Metaverse content in front of our eyes and blended into our reality. The technology for this vision to become the reality that we need is here, or it will be available in the near future, and this is a fact as its development is fuelled by billions invested by companies around the world. Everything is converging. The next wave of technological transformation is building. The Metaverse is coming.

Endnotes

1. Statista (2021) 'Number of mobile devices worldwide from 2020 to 2025 (in billions)', 24 September, available at: https://www.statista.com/statistics/245501/multiple-mobile-device-ownership-worldwide/ World population is at approximately 8 billion at the time of writing.
2. Neufeld D (2021) 'Waves of change: Understanding the driving force of innovation cycles', *World Economic Forum*, 5 July, available at: https://www.weforum.org/agenda/2021/07/this-is-a-visualization-of-the-history-of-innovation-cycles/

CHAPTER 2
DIGITAL PRIMITIVES AND EMERGENCE – A CASE FOR DIGITAL SELECTION

As I write this, we are at the initial phase of Metaverse adoption, in the stage that I describe as a digital primitives era. The digital primitives are the initial use cases for a new technology or medium and are closely linked to existing use cases that appear in other pre-existing forms.

A good example is the websites created in the 1990s, which were little more than glorified brochures. They were very limited in terms of the value they provided, especially in comparison to the use cases for the Internet at present. Using the Internet, today, you can buy goods, plan a holiday, find love, hail a cab, order food and even find out what song is playing in the background. All of these activities were technically possible in the early days of the Internet, but nobody realised them, because some of the technical requirements were not yet matched by the technology of the time, or simply because nobody had realised these were valuable use cases for this technology.

As I explained in the previous chapter, use cases in any new medium also tend to imitate the look and feel of use cases available on previously existing mediums or technologies – a phenomenon also known as *skeuomorphism*. Do you remember when the iPhone was first released, and its Notepad app looked exactly like a notepad? That's a perfect example of skeuomorphic design, because they based their design for this new medium entirely on an old medium. While this isn't necessarily bad practice, and in many cases can be a natural way to introduce users to a new medium, it may not necessarily be the best way to take full advantage of that new medium.

When it comes to the Metaverse, we are currently in the midst of this digital primitives era. We're scratching the surface of what's possible, because we're still trying to work out what the Metaverse could be and what we can do with the technology that's behind it. At the moment, we can see some rather innovative use cases for blockchain, virtual worlds and non-fungible tokens (NFTs), but what use cases will we discover in the future? How will these evolve as we develop interconnectivity between these worlds, portability of these worlds and as digital content starts to bleed into the real world? What kind of use cases will be possible then?

During this digital primitives era, various use cases will bubble up, and, through a process of natural selection directly related to the business generated, the truly transformational use cases that will also be the most successful business-wise will emerge. This is not only true of the technology itself, but also of businesses that are popping up to use that technology. Innumerable start-ups are created during every wave of technology. We saw it happen with the Internet, mobile technology and, even as I write this, it's happening again with the Metaverse. Every time, only a few of these start-ups go on to become truly successful businesses. Why? Because there are certain use cases that are more refined than others. This leads to what's known as *emergence*, as I explained in the first chapter.

Natural selection in a digital era

Emergence is when one or multiple use cases become considerably more successful than the other use cases that are around at the same time. This leads to a form of natural selection within the digital world. Emergent use cases for mobile devices include social media, the gig economy for taxi rides and streaming services for music and movies.

We don't know what the emergent use cases will be for the Metaverse yet. We are very much still at the digital primitives stage when it comes to the Metaverse, and we haven't really begun to explore its full potential. Take NFTs as an example. An NFT is a digital certificate of ownership. It's a smart contract that certifies that a user owns specific rights, a specific item or specific intellectual property (IP). Therefore, attaching a jpeg or piece of art to an NFT sounds to me very much like the basic websites we saw being developed in the 1990s. NFTs, for jpeg art, are a digital primitive. They are the first use case of this new technology, but, as we will discover, this has far more potential.

Generally, it takes between 5 and 10 years to understand the potential of new technology, the right user experience and the right business model. This is the point of emergence I talked about in the opening chapter of this book – technology, user experience and business need to converge in order to create a transformation use case. This is the real recipe for success.

To come back to NFTs, at the time of this writing, it's still cumbersome to create an NFT wallet, and the technology still feels very clunky. There is friction around the access to the technology required. The business model is, at best, still rather cloudy and fluffy. There are a lot of questions that remain – what value does purchasing an image attached to an NFT really have?

For example, the Bored Apes Yacht Club (an NFT collectable series living on the Ethereum blockchain) released NFTs in 2021. If you purchased one, you gained access to certain events, and you became part of an exclusive club of people who could put their Ape as their profile picture on Twitter. But what did that really achieve? It allowed you to show off that you purchased a Bored Ape when they sold them, but this really makes the NFT little more than a status symbol. It's no different from buying and wearing a Rolex. We are still waiting for the transformational use case to emerge, although we'll explore some potentially ground breaking use cases for NFTs as we move through this book (there is much more to them than Twitter profile pictures, I assure you!).

Emergence is a process that happens naturally as we experiment with new technology and new mediums. In the early 2000s, for instance, there was an invasion of reality TV shows, starting with the likes of *Big Brother* and then moving on to *Love Island* and *The X-Factor*. In the 2020s, we are seeing these programmes dying out and new waves of shows emerging instead, such as the multitude of documentaries on streaming services covering everything from true crime to the natural world – move over Simon Cowell, the conversations are about Joe Exotic now!

It is a form of natural selection, which is based in part on trends, as well as on the people who are using a specific medium, how people's values align to a specific medium and how society views that medium – but it is also directly linked to measurable business results.

When we're talking about a specific medium, the target demographic is also very important. For example, the target demographic for linear TV is people aged 55–60 plus. The majority of younger adults watch shows on streaming services like Netflix, Disney+, Amazon Prime or YouTube. Teenagers are more focused on watching content via social media such as Snapchat, TikTok and Instagram. Emerging use cases are therefore influenced by the success they find with specific demographics within society.

The Metaverse will be a new infrastructure, and this will lead to the emergence of new use cases that spawn directly from that infrastructure and the platforms that operate within it. You could say that the way in which it will evolve is much like natural, biological genetic mutation and adaptation. Just as we see, in humans, greater diversity in populations where there is a bigger and more diverse gene pool, so too, with technology, we will see more varied use cases evolve when we are working with a broader pool of technological innovations. Sooner or later, some of the use cases that emerge will out-compete others and become more successful.

Eventually, we will see those successful ideas intertwining. We've already seen this in certain areas of mobile tech, for example, with Uber and now Uber Eats developing. Where in the digital primitive stage we were defining use cases based on pre-existing mediums, once we reach the emergence stage, we start associating new use cases with those that have already been proven to be successful. When you can anchor an idea to another one that is already emerging, your idea has a greater chance of succeeding.

From the familiar to the impossible

There are two main reasons why we often start with skeuomorphic design in the world of technology. The first is because this leads to the creation of something familiar to users, which makes it easier to use and understand in that technology's early iterations. Let's look again at the notepad apps on our mobile phones. The first notepad apps looked like notepads and, as a user, I understood what I could do with it (make notes). When I made text notes, they even appeared in a font resembling handwriting. From a usability perspective, this made it familiar and therefore less off-putting to new users. However, did this type of notepad app deliver the best user experience? That's unlikely.

Another great example is the evolution of mobile website design. The first mobile websites were very similar to desktop sites. Then developers realised that the user experience for mobiles needed to be optimised for touch devices – this is when 'burger' menus started appearing on the top left or top right of websites, making it easier to navigate using your fingers and thumbs. Other small features, like the back button, got added gradually.

When you stick too rigidly to a skeuomorphic design that mimics something in the real world, you aren't able to take full advantage of what the technology can do. In the case of mobile devices, we've seen notepad apps evolve to better utilise the technology behind them. Now, when we open a notepad app, we don't see a lined page but a blank slate. The interface is optimised for touch, allowing us to easily change the font, colour and so on. The apps have evolved.

In both these examples, there was no immediate shift from skeuomorphic design to native design; just like the emergence of use cases for new technologies, native designs gradually emerge and evolve to deliver an improved user experience as that user experience also evolves. Within the world of mobile app design, some have definitely been more transformational than others. *Flipboard* has been a revolution in the field of tablet and mobile readers, for example. This app downloads news content into pages optimised for mobiles and allows you to 'flip' from one article to the next using your thumb. When *Flipboard* first came out in 2010, it was truly revolutionary in comparison to the other RSS feed aggregators available.

This is a perfectly natural process for new technology to follow. The second reason for skeuomorphic design is that new technology is usually invented to deliver an initial use case or solve a specific problem. The initial use case (or cases) for any new technology is generally linked to previous technologies and/or previous use cases. For example, blockchain technology is great for supply chain management because it allows you to understand the origin of specific assets at any point in the supply chain. It's ideal for proof of ownership and proof of transactions, which is why it's become closely associated with finance and cryptocurrencies.

However, there are other interesting use cases for blockchain, as we'll cover in greater detail in Chapter 5. For now, as an example, let's consider how this technology could be used to allow individuals to manage the ownership of their data, where they are able to decide who should have access to what data, who should own their data and whether they need to revoke access to their data at any point in time. This was not one of the original use cases for blockchain, but that does not mean that it's not a valid and potentially fantastic one.

So, although new technology may be designed with a specific use case in mind, once it is out in the world, it has the potential to evolve. New use cases bubble up, and this can lead to the discovery of transformational use cases that have much wider impacts on how we live our lives.

Familiarity is a very important box for new technology to tick, though, because this allows people to feel comfortable as they become familiar with a new medium and environment. Ask yourself why we have trees in virtual 3D worlds? Why don't we have worlds that are completely fictional, where the laws of gravity don't apply? The reason is that we have to go on a journey that starts in the familiar before we can get to the impossible. We have to begin with something that is similar to the real world before we can start to explore possibilities that are impossible in real life and more distant from the real world.

We're entering an age of exploration

The age of exploration of the potential in 3D worlds and immersive experiences has already begun. For example, while many virtual reality (VR) games created to date have been the likes of first-person shooters, there are some that are breaking the mould. One that stands out for me at the time of this writing is *Echo VR*, which you can play on the Meta Quest VR headset. If you've seen the film *Ender's Game*, you will likely remember the scene where they play handball in space. *Echo VR* is similar, except the ball is swapped for a frisbee. It's essentially a multiplayer game staged in a huge arena, where you float in this immersive world, playing alongside your teammates.

I believe this is just the beginning of the exploration of these kinds of experiences that are completely different to our physical reality. We are going to move towards a virtual world more similar to that depicted in the book *Ready Player One* as we begin to understand the true value of creating a different breed of experiences that are impossible to have in real life. However, in order to reach this stage, we have to begin by introducing people to virtual worlds that they find familiar and by helping them get used to them in a comfortable and safe way.

When the tech evolves to delivering hyper-realistic immersive experiences, we won't simply be able to drop people who have never even worn a VR headset into the middle of space and expect them to play space frisbee. It's important to guide this new generation of users into virtual worlds by the hand, initially starting with virtual worlds that look and feel more familiar before making virtual worlds that look and feel more alien, and then branching out from there.

This is a completely normal process to go through, because people need to learn how to use this new technology. Once we have people who are Metaverse natives in our midst, our artists, creators, software houses and platforms will have the freedom to open up their creativity – and, in doing so, different use cases will emerge.

What's in it for us?

Clearly explaining the benefits of adopting any new technology is going to be key to its success and the rate at which it is adopted by the wider society. When it comes to the Metaverse, there are a multitude of benefits, even in its early iterations. In essence, the Metaverse is about experiencing immersive and engaging realities with other users, so if you can create a digital reality that is similar to the real world, you can use this technology to shift some of our real-world behaviour.

Telepresence is a good example, because this gives you the opportunity to attend events all over the world without the need to travel. You can attend events with your friends or work colleagues. You could reduce your business' need for real estate by creating a virtual office where you can interact with your team. These are examples of how you can use technology to model what is happening in your real life right now.

However, the benefits of adopting this kind of technology go far beyond what's possible in the real world. If we stretch the spectrum of reality, we can find all kinds of new possibilities to use this technology for incredible benefits. Let's take the concept of having a virtual office and stretch it to creating a 3D representation of a tumour, so that a surgeon can step into the brain of the person they need to operate on, and manipulate that tumour and see details that would previously have been impossible to discern. This might sound like something out of a sci-fi novel, but it is already happening. A friend of mine had surgery on a brain tumour, and the surgeon operated using a Microsoft HoloLens mixed reality headset, so that a 3D model of the tumour was projected onto my friend's skull. This meant the surgeon could see the exact location of the tumour in relation to the position of my friend's skull as they were operating.

Our journey of discovery is just beginning

For many people, the concept of the Metaverse sounds rather fluffy. There is a belief that there isn't a solid use case for the Metaverse, beyond perhaps online gaming. Many people don't feel as though this is going to change our lives, because the kinds of use cases we're seeing right now – with the likes of buying NFTs, cryptocurrencies, digital worlds, digital clothes and so on – do feel fluffy, and they don't feel 'real'.

However, as I said in the Introduction, we are only at the beginning of our journey of discovery, and we have no idea where this journey is going to take us. We are on the edge of a greyed-out map, but as you explore further into this world, you find amazing things, including things you didn't expect.

As the map of the Metaverse appears completely grey to us, we need to start exploring. As with the game *Breath of the Wild*, the people and companies who start exploring these technologies first will be the ones who finish the game first. They will be ahead of the pack, because they will gain experience, knowledge and technical know-how about how to design and build new elements within this digital world.

This is one of the major reasons why there is a lot of money being poured into the Metaverse at the time of this writing, because businesses are acknowledging that this is a journey we need to go on, even if we don't know where it's going to take us yet. Sony has invested $500 million in *Epic Games*, the game publisher behind the game *Fortnite*; the consulting giant Accenture opened a Metaverse Continuum Business Group, which is a new pillar within the company that is focusing entirely on the Metaverse; and Mark Zuckerberg has described the Metaverse as the next big technological leap after mobile Internet, changing the name of his company from *Facebook* to *Meta*.

We are at the beginning of a new era of communication, commerce and experience. It is a huge opportunity, because it's not only about impacting real life, but also about creating parallel lives and parallel experiences. The beauty of being a gamer is that you can live a thousand lives. You can live the life of Link from *Zelda*, or Kratos from *God of War*, or Master Chief from the *Halo* series. Right now, living these lives is a very personal, and at times surreal, experience, because you do it in the context of these games. In the future, however, you will be able to live multiple lives, with multiple identities, in alternative versions of the real world, socialising with your friends and family, going to events and even working as another version of yourself. In taking this leap, we will be expanding our horizons of what we can learn, do and explore.

As a society, we don't yet know where this is going to take us or what transformational use cases will emerge. But this is going to be massive. We have only just started exploring the 'map', and we don't know what we'll find out as we venture further into this new realm. We are the pioneers of the Metaverse.

CHAPTER 3
FROM DYSTOPIA TO UTOPIA – ENTER THE METAVERSE

The concept of the Metaverse originated in a science fiction novel originally published in 1992: *Snow Crash* by Neal Stephenson. This novel very much veers towards the dystopian idea of a Metaverse, as have many other representations that have followed it. In the book, the Metaverse is a place where people are finding shelter from a dying world. You could draw some parallels with the world we live in now, where we're facing a global environmental crisis, but many of those who are involved in developing the Metaverse as I write this are excited by the utopian vision for the Metaverse.

So, what does a utopian Metaverse look like? It's a place where people can socialise, discover, learn, build businesses, connect with their loved ones and make memories that are impossible to create in real life. You could play tennis on the moon, or go on a *Star Wars* adventure with your friends. You could also tap into some incredible social and professional opportunities. Imagine learning to be a chef without ever stepping into a kitchen? Or think of the possibilities it offers for people in remote locations, who could learn to use everything from engineering to medical equipment that is otherwise inaccessible to them. It is a virtual land of opportunity.

The Metaverse is a place where you can create and live those stories, both on your own and with your friends. In the future, people are not just going to live one life; they will be able to live multiple, parallel lives in different universes and even have different identities. The phrase 'you can be whatever or whoever you want' takes on a whole new meaning in the Metaverse – you can be a businessperson, professional athlete or doctor. You can be male–female or non-binary. You can be a robot, a fairy or even a unicorn if you feel like it! How about a non-binary unicorn doctor? No judgement here! Whoever you decide to be, you will be able to live those experiences and create those stories in whatever way you want to. The Metaverse offers us a new way to express ourselves, connect with others and experience different things, free of judgement, prejudice and all the other social restrictions or barriers the physical world presents. It's true freedom, and this is the beautiful aspect of the Metaverse.

Of course, there are some dangers, which I will explore in the coming chapters, but allow yourself for just one moment to consider the possibilities in this beautiful utopia.

Imagine a chemistry lesson. There's no teacher at the front of the class, no dry formulas scrawled in chalk on the blackboard. Instead, you are watching a chemical reaction occur in incredible detail. You can see the particles interacting with one another and you are able to understand the process behind this specific chemical reaction. This is an experiential way of receiving information and one that will be made possible through the Metaverse. You'll be able to walk into 3D labs to learn about physics concepts, chemical reactions, biological processes – this is only the beginning.

Within the Metaverse, we will move from storytelling to story living. This means putting each of us as individuals at the centre of the story and making us believe in it. This approach is not only valid for entertainment, but also for education and learning, as well as any experience you care to imagine. As this book will show you, the worlds of entertainment, gaming, learning, education and life in general will begin to blend together into something truly amazing.

A utopian Metaverse = an inclusive direct democracy

To create this utopian vision of the Metaverse, it needs to be owned and maintained by the users themselves. It has the potential to be the perfect example of a direct democracy. If we look at a platform like *Decentraland*, we can see how this could work, with the landowners being afforded voting rights that are representative of how much land they own. This, in turn, allows them to steer the direction of the company.

Decentraland: *A decentralised Metaverse platform powered by the Ethereum blockchain where users can create, experience and monetise their content and applications. The 3D, traversable virtual space within the platform is called* LAND *and can be purchased (and then permanently owned by the user) using the platform's cryptocurrency called* MANA, *which can also be used to purchase digital items on the platform that are sold as NFTs.*

This is a vision of how the Metaverse can develop and, in fact, a model for how companies comprising the Metaverse can be structured in the future. This also ties in with the vision for Web3 (as defined in 2014 by Gavin Wood, co-founder of Ethereum), which is a decentralised Internet where everybody can 'own' a piece of the Internet and, based on what specifically you own (whether a piece of code, a piece of 'land' or digital assets), you will be afforded specific rights. This is also vital for the shift towards a self-sustaining creative and digital economy, which I will discuss in much greater detail in Chapter 5.

This is the Metaverse we want to create; one where people are rewarded for their creativity and effort, and one where everyone is able to build on each other's work to lift the boats for everybody in a positive way.

A dystopian Metaverse = mass manipulation and control

As with any new technology, there will always be those who want to use it for nefarious purposes. Some of the biggest risks with the Metaverse are that it could be used for mass manipulation, control, indoctrination and data gathering, with this data then used to identify trigger points for sales and advertising material to exploit.

The beautiful movie *Ex Machina*, written and directed by Alex Garland, shows us an example of what a dystopian Metaverse could look like; one where an artificial intelligence (AI) agent inside a humanoid robot is trained based on your social media profile to find the best way to seduce you in order to obtain whatever it wants from you. This might all sound rather scary and make you want to run a mile, but the Metaverse does not have to go down this route. There are a number of rules and regulations that need to be put in place now to avoid this kind of misuse of the Metaverse and to keep both us and our data safe.

We need to make sure these rules and regulations are put in place sooner rather than later. Look at the next generation of virtual reality (VR) headsets that are being developed and prepared for launch as I write this book – they include features like face, eye and body tracking cameras, which means they will be able to not only track what you are looking at, but also your microexpressions, the dilation of your pupils, your body posture, your physical fitness and even your heart rate using micro variations in your skin colour. In addition, they could potentially access the recordings of everything you say in the Metaverse. This is a massive amount of data that, if not properly protected, could easily be used to model tailormade algorithms and AI agents designed to trigger an emotional reaction or action by any user.

Due to the immersive nature of the Metaverse, we will be sharing a huge amount of personal data whenever we are present on one of these platforms. Your behaviour in the Metaverse creates a digital data lake and, just like swimming in a real lake, every movement you make moves the particles around you. When you're fully immersed in the water, you're constantly creating turbulence around your body. Similarly, when you're fully immersed in digital content, every movement you make will create digital data, which can be fed into machine learning algorithms and used for all manner of purposes, including to create very realistic and believable virtual human avatars, which I'll discuss in more detail in Chapter 9.

We will explore the risks, as well as the steps we can and need to take to reduce them, later in the book. As the Metaverse is developing, we have to consider these concerns and take action to prevent such activity to steer it towards a utopian Metaverse, rather than a dystopian one.

The blueprint for the Metaverse

We want to work towards creating the utopian vision of the Metaverse, but what does that look like in practice? In this vision, the Metaverse is a series of interconnected 3D worlds, through which users can seamlessly move. As users move from one world to the next, they can not only bring their identity (the appearance of their avatar), but also virtual assets, whether that is a designer handbag, a pet or even a car. These virtual assets can be representative of items that each person has in the real world, through the use of non-fungible tokens (NFTs) and digital twins, which I'll discuss in greater detail in Chapter 5.

However, virtual assets can also bleed into the real world in other ways. Layers of digital information could be overlaid on top of your field of vision while you're walking down the street, for example. Potentially, this could enable you to overlay your Metaverse avatar onto your physical self when you are in the real world for others wearing augmented reality (AR) glasses to see, or it may be something more subtle, like overlaying your digital sneakers on your shoes, again for others with AR glasses to see while you are out and about. The idea is that the Metaverse will become an interoperable ecosystem of platforms that allow an economy to evolve that features products that are portable between these various platforms.

The ultimate aim is for people to be able to move between virtual platforms and the real world in a seamless and natural way. For this to occur, the enabling technology and its infrastructure needs to be almost invisible to the user, the interface needs to be as natural and immediately intuitive to master, and access to services should be as frictionless as possible. I'm talking about having the ability to interact with all of these different worlds without the need for menu diving or complex interfaces. One of the keys will be developing a frictionless system that enables payments across different platforms, as well as allowing us to move the purchased digital goods from one platform to another.

This frictionless system is essential for the growth of the Metaverse, because the more difficult it is to purchase or use assets or to access services, the less success those services, assets and places will have with the wider public.

There are two strands we need to be aware of as the Metaverse develops – the components that make it appear in a certain way, and the components that make it work in a certain way. The components that deal with the appearance of the Metaverse relate to the experience design – which is the mechanics that makes users come back to a certain Metaverse world which is at the heart of the entire Metaverse experience – and to the topology of both the virtual world and the real world where digital content is overlaid. This essentially relates to how the virtual lands are designed, and, within that, there can be different layouts and visualisations of the Metaverse.

Alongside these components are also the ones that deal with the functionality of the Metaverse. These are the components that make the Metaverse work, whether we are talking about self-expression, the digital economy, moderation or digital assets. This group of components also incorporates all the hardware and software required to access the Metaverse, whether that be headsets, mobile phones, laptops or tablets. *How you access the Metaverse shouldn't matter, because it should be available from any of these devices. However, the hardware you use to access the Metaverse will determine how immersive your experience is.*

As I write this, there are haptic suits that deliver force feedback, so if you are playing a video game in VR and get hit, you will physically feel it. Obviously, if you are wearing a suit like this, your level of immersion will be far higher than someone who is playing the same video game on a PC. High levels of immersion won't appeal to everyone in every situation either. A couple of years ago, I started playing the famous horror game *Resident Evil 7* on my PlayStation VR headset, and had to stop because I found the experience too intense (it genuinely scared me!). Even though today's VR capabilities are still pretty basic, it was too much for me, so I stopped playing in VR and finished the game on a normal TV screen.

Reading this, you might be envisioning a Metaverse that is something akin to what is described in *Ready Player One*, but we are still a long way from achieving that level of sophistication in virtual worlds (I predict at least 10–15 years to reach that level, if not more). Concepts like the omni-directional treadmill, moving seamlessly from one world to another, the level of graphics and interaction, as well as all the non-playing characters that are animated by AI – all require a level of technology and sophistication that, in 2022, we simply don't have.

We don't even have the computational power required to run a Metaverse of this kind at the time of this writing. Intel executive Raja Koduri has even said that the computational power required to run the future Metaverse infrastructure will need to be 1,000 times higher than what is in existence right now[1]. We still have limits on multiplayer games for precisely this reason. In fact, even some of the most sophisticated multiplayer games currently on the market, can only have up to a few hundreds concurrent players in a single game instance (able to see and interact with each other). In the Metaverse, you will potentially want to have millions of people in the same virtual space at the same time. Right now, this is impossible.

Limitations are imposed by our networking, computational power and protocols of communications between servers; these are all challenges that will need to be addressed before we can even begin to come close to the kind of all-encompassing Metaverse showcased in *Ready Player One* and deliver a credible experience for users. While organisations such as Improbable (www.improbable.io) are working on improving networking technology and even running advanced military simulations that simulate hundreds of thousands of users on the battlefield at the same time, we have a lot of ground to make up before this becomes viable outside of specialised applications.

The current status of the Metaverse

How far are we from this vision of the Metaverse, where a series of interconnected digital worlds bleed into the real world, with their own economy and millions of users able to participate in the experiences, games and learning opportunities? This highly experiential world, where lines between the real and the virtual blur, is in sight, but some way out of reach, based on our current technological capabilities, as I have just explained.

In 2022, as I write this, the Metaverse is currently made up of various Metaverse platforms, each with their own collection of 3D worlds that are fairly basic in terms of look and feel when compared to modern AAA videogames, and these are nor interconnected with each other. The likes of *Decentraland* and *Roblox* are best described as self-contained community platforms. One of the barriers to interconnectivity at this stage is that there are no clear standards of interoperability or networking between those different platforms.

> ***Roblox:*** *Roblox is an online gaming platform and game creation system. It's home to a global community of millions, and where developers and creators use the tool* Roblox Studio *to produce their own immersive multiplayer experiences.*

It's important to note that the Metaverse is not just the user-accessible layer of interactive platforms. Rather, the Metaverse is the full infrastructure that underpins these platforms, connects them, and enables the whole mechanics of authentication, digital identity, transactions, ownership, and content distribution – just as the Internet is the full infrastructure that underpins the various sites, platforms and content that you can find online and not just the superficial layer we see on our screens.

The fragmentation of standards, platforms, economies and currencies is one of the greatest challenges that will need to be overcome if we are going to move towards the kind of Metaverse I described at the beginning of this chapter. There are multiple organisations that are already working on solving these issues and bringing the Metaverse closer to becoming a reality.

One example of these organisations is the Open Metaverse Alliance, which is working on creating rules and standards that can be applied across platforms in the Metaverse to solve some of these key interoperability issues. The Khronos Group is also working on developing interoperability standards specifically for 3D assets within the Metaverse. Their work will be important for allowing digital assets to seamlessly move across platforms within the Metaverse, and you'll learn why this is so important for our experience as users as we move through this book.

The Khronos Group has also created a standard known as *OpenXR*, which will be very important for those creating and building 3D experiences that are usable across multiple devices and multiple platforms. It is a high-performance, low-level application programming interface (API) that is able to connect across devices (mobiles, tablets, VR or AR headsets, PCs, etc.) and platforms. This is the kind of technology we need to develop if we want to be able to seamlessly jump from one Metaverse hardware to another. While consortia like the Khronos Group and the Open Metaverse Alliance are paving the way for the introduction of such standards, at the time of this writing, we don't have these standards, and it is very likely that they won't be fully designed, agreed, and broadly adopted before the next three to five years. We are, therefore, unable to experience the current Metaverse platforms as interconnected and interoperable.

Therefore, the best way to describe the current status of the Metaverse is as a series of fragmented platforms that are each doing their best to create their own self-sustaining economy and community. They are almost like small frontier towns in the Wild West before the arrival of the railroad (and you could argue that the Metaverse is a bit lawless like the Wild West was back in the day). All these Wild West towns were self-sustaining and governed by their own laws, living off their own patches of land. Then the railroad arrived, introducing a means to travel quickly and easily from town to town. It opened up trade, brought communities closer and brought about the standardisation of the law. We are still laying the tracks of the railroad that will connect the Metaverse platforms, but the train has definitely left the station.

Who will own the Metaverse?

It is very important to stress that the Metaverse will not be owned by any major corporation. There has been a great deal of confusion around this as a result of Facebook's rebranding to 'Meta' in 2021. One popular theory is that the Metaverse will be owned by the community, where the 'land' owners create the rules and regulations within the Metaverse, especially if decentralised platforms become more prominent, as this is their operating model.

It is also important to make a distinction between centralised and decentralised platforms. Meta's *Horizon Worlds* and Roblox are examples of centralised Metaverse platforms that work with a central server and have a board of directors who direct both the company and the rules of the worlds they support. On the other hand, we have decentralised platforms like *Decentraland*, where everyone who owns 'land' on that platform will have a say on its rules and how it is running, as I explained earlier.

Horizon Worlds: *A free online VR video game and game creation platform owned by Meta. It is described as a 'social experience', where users have the ability to create a wide variety of immersive experiences.*

The Metaverse will be a constellation of all these platforms; some will be centralised, and others will be decentralised. The centralised platforms will be owned by private or publicly listed companies (like Meta), whereas the decentralised platforms will be owned by the platform users, and will have a completely flat management structure that relates to the ownership of the platform itself, also known as decentralised autonomous organisations (DAOs). This means the Metaverse as a whole will not be 'owned' by anyone, and it certainly won't be owned by one company.

As a result, it is very important that we develop these common standards and rules that will apply across platforms within the Metaverse, because otherwise it is going to become a true Wild West. We will need to create entities that are vigilant about the content on the Metaverse and that support the regulation and governance of the Metaverse to ensure all our safety – not only of those who inhabit the Metaverse, but also those in the real world – because the boundaries between the two will become increasingly blurred.

We have seen what can happen when organisations raised from the rampant adoption of a new emergent use case based on a new technology are left without effective governance or regulation – just look at the scandal surrounding Cambridge Analytica and their meddling in the 2016 US election and the 2013 Brexit referendum. This is why it is so important that these regulations be put in place – and, the earlier we can do this, the better. *Some organisations have already started defining the guidelines for the Metaverse – for example, the World Economic Forum has created a committee of privately held businesses, subject matter experts, academia and public-sector policymakers aimed at creating the general guidelines for an economically viable, accessible and inclusive Metaverse.* Throughout the rest of this book, I will discuss these regulations and the building blocks we require in different areas to create a utopian Metaverse that allows us to live our best lives and stay safe while doing so.

Breathing life into the Metaverse

As you have seen, the Metaverse is still in its nascent stages, and there are a number of core pillars we need to have in place to breathe life into the Metaverse and allow it to realise its potential. Briefly, these are:

- An easily accessible and secure digital economy.

- Interoperable, interconnected and persistent 3D worlds and platforms that will operate according to specific standards and regulations.

- A way to host a high number of simultaneous users across all the different platforms to allow them to share social experiences.

- A seamless way to access this world from any device, from the lowest entry point (like a mobile phone) or top-tier hardware (like haptic suits and high-end VR headsets).

- An infrastructure that provides hardware-agnostic access.

- A network infrastructure that has the capability to sustain all of this activity.

- Well-structured and widely adopted governance paradigms that will cover user safety, privacy, content moderation, sustainability and responsible AI to ensure that the Metaverse does not become a dystopian environment; these paradigms should be produced in orchestration with the proper entities (such as national governments and supranational organisations like the UN), rather than being dictated by the companies that develop platforms in the Metaverse.

Throughout the rest of this book, I will explore each of these topics in more detail. The main message, particularly when it comes to governance in the Metaverse, is to learn from the mistakes that have been made in relation to the regulation of new technology and social platforms in the past. We should not be scared of this new technological revolution, because the opportunities it offers us are exciting and expansive.

While the Metaverse might sound like something new, the components that will underpin it, such as blockchain and multiuser technology, have been in development for many years. As I write this, we have reached a critical mass, an inflection point where the curve of user adoption is growing exponentially. As a result, we need to act quickly to ensure that users feel safe when accessing the platforms that make up the Metaverse, both now and in the future. The gaming industry has long been at the cutting edge of technological developments, and it is one of the essential building blocks for the Metaverse. It's time to suit up and start exploring the Metaverse map.

Endnote

1. *Business Insider India* (2021) 'Metaverse vision requires 1000x more computational power, says Intel', 16 December, available at: https://www.businessinsider.in/tech/news/metaverse-vision-requires-1000x-more-computational-power-intel/articleshow/88316064.cms#:~:text=Metaverse%20vision%20requires%201000x%20more%20computational%20power%2C%20says%20Intel,-Advertisement&text=Leading%20chip%2Dmaker%20Intel%20has,from%20what%20we%20have%20today

CHAPTER 4
FROM GAMING TO METAVERSE – HOW GAME MECHANICS AND TECHNOLOGY ARE GOING TO CHANGE THE WORLD

With contributions from Bob Gerard

Bob Gerard is Learning Lead for Accenture's Metaverse Continuum business group, and also leads the organisation's Learning Ingenuity group. In his role, Bob focuses on helping everyone at Accenture learn better. Within the Metaverse Continuum business group, his role is to help people develop the skills they need to work within the Metaverse as it grows. He also describes himself as 'a child of the videogame era' and has a lifelong passion for gaming.

When it comes right down to it, humans learn best through stories. We have been storytelling for centuries; there's a reason that Aesop, the Brothers Grimm, and most spiritual leaders over the millennia have taught through stories, instead of through PowerPoint presentations!

We started with pictograms painted on cave walls, and now we've evolved our storytelling to include immersive virtual reality (VR) experiences. While language has been a very powerful tool for us, visual communication is a particularly powerful way to tell a story. As individuals and societies, we have always been very attracted to a good story. One of the other reasons why we love listening to stories so much is that they allow us to envision other possibilities for our lives.

Storytellers are entertainers, but, more than this, the people who can tell good, powerful stories also perform well in business. If you look at Steve Jobs, you can see that part of his success with Apple was due to his fantastic storytelling abilities. He used the power of storytelling to get excited and bring them along with him.

One particular aspect of storytelling that draws us in is the chance it offers us to live other lives while remaining in our own skin..., *pretending*. We can see this reflected in pictograms, books, radio and TV dramas, movies and, of course, video games. Immersive storytelling has always hooked humanity, and being part of these stories is deeply ingrained in us. Videogames, in particular, deliver an immersive experience like no other. People who don't play games only get to live one life, whereas gamers can live thousands of lives.

Gaming is the bridge that allows you to transition from hearing a story to actively participating in a story. This active participation is something we all crave. Just look at the way in which children develop and grow their understanding of the world; an important part of that developmental process is pretending. Children play games where they pretend to be different people and where they step into different roles. We learn by telling stories.

Children's brains have what's known as *neuroplasticity*, which is a characteristic that allows them to adapt to different situations very quickly. It's this characteristic that allows children to jump from being Captain Hook, to Luke Skywalker, to Superman – all in the space of a few hours. They can move from one world to another, adapt themselves, imagine themselves in different situations and act accordingly. They can create who they are by designing an imaginary world in their brains. However, this neuroplasticity gets lost over time.

As adults, we like to be in our comfort zone with what we know. We have our house, our job, our family – it's all very familiar. This means it's increasingly rare for us to find ourselves in situations that are not familiar or comfortable in some way. As adults, this makes it harder to adapt and to embody different characters in the way that children do.

In addition, pretending to be someone else, and even playing, gets discouraged as we get older. If you want to have the sensation of pretending to be somebody else and immersing yourself fully in someone else's story as an adult, you almost have to resort to videogames. The only other option is live action role-playing and cosplay, which will almost invariably get you branded a 'nerd' (although increasingly less so ... more on that later!And besides, there's nothing wrong with being a nerd.)

As we grow up, we become increasingly tied to the real world, where the rules are always the same. However, this gets thrown out of the window with videogames, because you can master a virtual universe by understanding, and potentially bending, the rules within it.

The rise of videogames

The availability of personal computers (PCs) during the 1980s was a significant factor in propelling videogaming to popularity. PCs were initially sold as a way to make people more productive in their work, but one of the first things people used those computers for, aside from work-based tasks, was for playing games.

'When I was 11, I asked my father for a computer. I told him that it was because I wanted to learn how to program, and, obviously, I asked for an Amiga 500, the best gaming machine on the market at that time – because, of course, I mainly wanted to play videogames. Instead, my dad bought me a PC XT 8086 from Olivetti with a terrible green phosphor monitor (definitely not a gaming machine) – at least this propelled me to self-teach how to program in BASIC to write some rudimentary games to play!'. Nick

Very quickly, gaming became a very important use case for PCs for many people. In fact, the vast majority of people who own a PC will have at least one game installed on that computer, often alongside the software they use for work.

Following the collapse of the home game console market in the early 1980s (culminating in the infamous burying of hundreds of Atari 2600 'ET' cartridges), PCs became the platform of choice for home videogames. We can remember how those of us who played videogames in the 1980s and 1990s would upgrade certain elements of our PCs to improve the experience, such as through AdLib and Creative Sound Blaster audio cards. We can even recall Sierra On-Line putting disclaimers on the boxes of its videogames, effectively warning us to not bother buying one of their games if we didn't have an audio card. At the time, many felt this was really bold, but game designers knew people wouldn't get the full experience without a sound card.

In the early 1990s, it was the launch of *Myst* that really helped the CD-ROM catch on, because you needed a CD-ROM if you wanted to play that game. This meant many people either added CD-ROMs to their computers or simply bought PCs that already had CD-ROMs installed. Our point really is that games have always been a powerful underlying force in terms of the uptake of new technology.

'My first computer was a TRS-80 Model 3. It had 4K of RAM, and there were a few games available that allowed you to move and manipulate characters. I remember playing Star Trek on my TRS-80. But the games that really grabbed my attention and riveted me were the text-based adventure games. These were some of the first computer-based role-playing games, where you actually had a character who grew. They completely absorbed me'. Bob

One of the reasons why many people, including Bob, were so captivated by text-based games was the ability to be fully immersed in the narrative. However, as technology progresses, there are many new layers that are being applied to create this truly immersive experience, and this kind of experience is not limited to videogames. It will form a key component of the Metaverse too.

Technology is catching up to the vision

Over the years, many organisations have experimented with creating their own virtual worlds, and these have often taken the form of learning environments. They might have been text-based environments, much like those early videogames, or incorporated live chat. There may have been a layer of pretending required by those participating, and this is where virtual worlds for activities like learning can fail.

The kinds of text-based worlds that gripped Bob relied on his engagement. All text-based virtual worlds require the participants to actively engage and assume their role; if they don't want to or aren't willing to do so, then the technology becomes cumbersome and a hindrance rather than a help. This is why virtual worlds that are used for applications beyond gaming have been limited at best.

However, when the mass adoption of VR headsets begins, there will be no friction when it comes to entering a virtual world. This is when building virtual worlds for learning will really take off, because the technology, price point and our creativity are all converging to make that possible.

Gaming is the segue to this evolution of virtual worlds. Gamers have always been willing to test the technology and see what it's capable of, from those early text-based adventures through to early iterations of virtual worlds like *Ultima Online* (where the worlds were more graphical, and people were able to create full economies), and now to some of the complex and engaging virtual worlds that have been created, such as the likes of *World of Warcraft*.

It was the evolution of graphics and interactivity that brought us the gem that is *World of Warcraft*. Not only has this game been groundbreaking in terms of its storytelling and interactivity, but also in relation to its community creation. People meet at conventions, they get married in *World of Warcraft*, they live entire lives within this virtual world. This is just one example of how those early iterations of technology, which weren't exactly easy to engage with, have evolved substantially over the last 20–25 years into interactive environments that facilitate huge levels of engagement. To engage in such rich adventures, gamers in the 1980s had to endure those early, clunky text-based adventure games, and, at times, it really was a labour of endurance!

However, the more the friction of joining those worlds was reduced and the fewer barriers there were to entry, the more mainstream gaming became. Part of this was due to the technology evolving to a point where such virtual worlds became much more engaging, but you also can't escape the ubiquity of social media. Social media platforms have given everyone a voice, and this has allowed conversations about virtual worlds and gaming as a whole to reach more people, while giving them a safe way to explore it before they decide to dive in.

If you look at the renaissance that *Dungeons & Dragons (D&D)* is having at the time of this writing, you can see how important this community engagement is. Arguably, *D&D* would not have returned to the level of popularity it's currently experiencing, were it not for the US web series *Critical Role*, where a group of voice actors play *D&D* and stream their campaign sessions. The series is broadcast on Twitch and YouTube, and these social media channels have enabled it to gain traction.

Social media has created the resin that binds these communities together. In the past, we had many tribes of gamers who were deeply disconnected from each other. Now, however, using social networks and the Internet, they have been able to connect with one another in a self-organised way outside of the boundaries of the games themselves.

The concept of communities and tribalism is key to understanding the Metaverse, because what it is trying to create is a deeply connected collection of communities and tribes. We are increasingly seeing gamers in the likes of *World of Warcraft* playing out actual life events within the game's universe. People have parties for their characters, they get married within the game and so on. They're not just playing the game, they're living a parallel life within this virtual world. The boundaries between games and real life are becoming increasingly blurred.

Identity and anonymity: from gaming to the Metaverse

One of the reasons people are drawn to gaming is the possibility it offers to live another life and to carve out a new identity. Identity will be fundamental to the Metaverse, and the world of gaming can provide a useful case study on how we create identities in virtual worlds and how we protect them.

In the most successful role-playing games (RPGs), you can choose every aspect of your character, from their race and gender to their clothes and accessories. We are already seeing something similar happening in Metaverse platforms. *Fortnite* is among the games that is bridging the gap between gaming and the Metaverse. Its creator, Epic Games, has made billions from microtransactions as a result of people making cosmetic customisations to their characters. These are not modifications that will affect or improve their performance in the game in any way. They are simply a way to express yourself and your identity. That might sound crazy to some, but it's really no different from paying to visit the hairdresser or getting a tattoo (and with a virtual avatar, at least you don't physically have to go under the needle!). Cosmetic customisations and the ultimate freedom of self-expression are features that will define the Metaverse.

However, as we have seen through online games like *League of Legends*, there is a balance that needs to be struck between building new identities, self-expression and anonymity. In case you're not familiar with it, *League of Legends* is a free-to-play, multiplayer battle arena game. The only thing the game's developers sell is the ability to customise your character, which means that you are basically paying to express yourself. You can buy more than clothing for your character too, such as dance moves and even the ability to dis the people that you kill.

However, the darker side of *League of Legends*, and gaming as a whole, is in the toxic online environments they can cultivate. *League of Legends* makes an interesting case study, with players in the game often treating each other poorly and being really nasty. Developer Riot Games has spent a great deal of time and money trying to combat the toxicity among its players, and it has had some success. One of the main factors that gives rise to such toxic behaviour is anonymity.

This is particularly important to consider in the context of the Metaverse. We should avoid anonymity in the Metaverse at all costs, because anonymity not only gives people a shield to hide behind when engaging in toxic behaviour, but it also enables people to pretend to be someone they're not. We're not talking about restricting people's ability to express themselves – if you want to appear in the Metaverse as a dragon because that's how you feel you best express yourself, then more power to you. What we are talking about is preventing people from creating avatars that pretend to be other people, or that are used as a platform to harass and degrade other users. We don't want to hand people a cloak of anonymity that emboldens them to bully, threaten or use the Metaverse for nefarious purposes.

In any given Metaverse gaming platform, despite appearing anonymous to each other, users should always be uniquely identifiable by the Metaverse platform provider in order to guarantee accountability for their actions.

Rules of engagement

Anonymity is not appropriate for the enterprise Metaverse, and it's not appropriate for a Metaverse where people engage socially rather than in a gamified environment. Those of us who game all know that we have only taken certain actions in a game, such as killing other people's characters, because we wore that cloak of anonymity.

We all have to know what the rules are, and we all have to opt into and abide by those roles, whether we are in the Metaverse or in a virtual game. For example, in a videogame, it might be socially acceptable to kill another character, but if we look at *World of Warcraft*, you have to opt into that experience. Before another character can kill yours, you have to opt into PvP (player versus player). When you do that, you know those are the rules.

One of the challenges we are going to have to navigate as the Metaverse grows is not only to distinguish where and when there might be different rules, but also to ensure that everybody who enters that virtual environment opts into those rules.

Ethics and identity

In early 2022, I was involved in a Metaverse social event with many people who were new to the Metaverse. They all chose their own avatars, and one man, let's call him Neil, selected a female avatar. We observed him being teased by two of the other men who attended, because Neil had chosen to appear as a female avatar in the Metaverse.

Now, Neil may have just wanted to try being in a different skin, but what if Neil was transgender and chose a female avatar to express his true self as a way of coming out to his friends? Had Neil's friends reacted in the same way as the two guys teasing him at that event, can you imagine how devastating that would be? This is why the ethics of the Metaverse are so important and, while it's still in its nascent stages of development, we need to consider such questions.

The solution to this particular situation has been to state that the code of business ethics that Neil and his colleagues follow in their organisation also applies in the Metaverse. Of course, policies surrounding the Metaverse are still evolving, and will be for some time, but this is a good starting point. If you wouldn't treat someone in a particular way in real life, don't treat them like that in the Metaverse.

We have to set an example and rein in any violent or threatening behaviour, even if we know what we're doing in the Metaverse won't physically hurt someone in the real world, because the Metaverse is going to become part of our reality, and it will be a place where we live part of our lives in the future.

There also needs to be a mechanism for reporting abusive behaviour, as well as consequences for that behaviour in a virtual world.

In a work environment, our Metaverse identity needs to match our persona and we need to be clearly identifiable. However, there may be a case for allowing anonymity in certain areas of the Metaverse, such as within virtual games.

Distinguishing gaming from the Metaverse

While gaming mechanics and technology have undoubtedly been, and will continue to be, powerful driving forces behind the creation and expansion of the Metaverse, it's important that we make a distinction between the Metaverse and any virtual games we play in the Metaverse, because certain behaviour will be allowed in a gaming environment that is not acceptable in society – and therefore not acceptable in the Metaverse.

For many people, playing extremely violent games like *Doom Eternal* is cathartic. It's an opportunity to relieve stress. In this particular example, you can project all your inner demons into this virtual world and fight them, which means you then don't bring those demons into your real life, into your interactions with your partner, children, friends or colleagues. This is incredibly important for many people around the world.

However, we have to make the distinction between games, which are pure entertainment, and the Metaverse, which is more like a parallel version of real life. That's not to say there may not be places within the Metaverse that offer pure entertainment, and where rules closer to those in videogames apply rather than those you'd expect in civilised society. However, such places need to be clearly labelled as such, so that we all know what to expect in different areas of the Metaverse.

We already have disclaimers on videogames, movies and even music about explicit content, abusive language and violence, and we believe a similar approach will need to be taken with any games or gaming environments within the Metaverse.

For example, there's currently a game called *Drunken Bar Fights* that you can play with the Meta Quest VR headset. It's a funny game that simulates a very violent bar fight. The game clearly is labelled for a mature audience, so you know that's what you're walking into, and you can be prepared for that.

What we have to get to grips with is the fact that the Metaverse will be a vast virtual environment, where we can move seamlessly from one place to another. With the click of a finger (or maybe a voice command), you can open and enter a portal into an entirely different world, but you need to be aware of the world that you're entering. If you were about to walk into a minefield in the real world, you would expect to (and want to!) see signs warning you that you were about to enter a minefield. We need to think about the Metaverse, and particularly gaming in the Metaverse, in a similar way.

Gaming: pretending within the rules

When you boil it down, gaming is simply pretending within the rules. If you watch two children playing in their backyard, and one picks up a stick and 'shoots' the other, the kid who has been 'shot' has two choices. They can say, 'No I'm not, you missed me!', or they could say, 'Okay, you got me, I'm dead', and then they will likely resurrect 10 seconds later. If the kid who's shot goes with the first option, the game probably won't last too long, because it's not much fun for either of them. However, if the kid who's shot chooses the second option, they have added a rule to that experience which they can then expand on to make a more complex game, even though they're still just pretending.

When you play games, you're pretending, but the rules have already been set for you. That said, you will still constantly be making decisions about your level of engagement.

"Star Wars: Galaxy's Edge" at Disneyland and Walt Disney World is an excellent example of this. It's not just a theme park land, but the best environment for 'pretending' Star Wars that there is. All of a sudden, the big cardboard box you had in the garden that you pretended was the Millennium Falcon has been replaced by the actual Millennium Falcon; you're not talking to the tall kid from down the block who always used to be Chewbacca, because you're now face-to-face with the real Chewbacca. You get to choose: do you embrace this and pretend that you're part of this universe, or do you treat it like a theme park and just say you want to go on the rides? If you pick the first option, that's when you really have fun, because you're interacting with this whole universe that's been laid out in front of you. If you want an even deeper level of immersion, you can stay on the "Star Wars Galactic Starcruiser" and take a simulated cruise through the Star Wars galaxy where you become the hero of your own story. It's a premium experience... with a premium price tag. But the depth of the immersion and the sense of truly being a character in Star Wars is so compelling, some fans have returned for multiple voyages on the "CSL Halcyon."

What Disneyland has created at Galaxy's Edge and the Starcruiser is one of the world's best immersive experiences. They elevate your imagination to such a high level that it not only becomes easy to engage with, but you want to engage with it. You feel passionate and excited about entering that world and being a part of it, rather than just watching from the sidelines. The point here is that immersion is a two-way street; if you don't engage with it and commit to the fantasy of it, it won't work. If you're resistant, you won't have a truly immersive experience. There will be friction.

The decision to be immersed in a virtual world isn't a one-time decision either. You have to make many micro-decisions that will continue your immersion. Will you stay in character with every interaction? In many ways, the effectiveness of immersion borrows heavily from improvisational theater, which is completely dependent on multiple people agreeing to stay "in character and in the moment." When you hear an improv person say "yes and," this is what they are talking about. Artistically, you can create an experience that it's easy to engage with, but it is still always each person's responsibility to decide whether they are going to stay engaged with and immersed in the world they see before them and continue to live the experience or not. The rules of the Metaverse are still being mapped out. We have the power to shape these experiences in our digital environments, just like we had the power to decide whether we got 'shot' when we were kids playing with sticks in the backyard with our best friends.

Playing to learn

We learn by playing. Playing allows us to create sticky memories, and this makes it the most effective way for us to learn new things. This is why there is a great deal of discussion around games for learning, because learning through play is more fun, more engaging, more memorable and more immersive. In short, learning through play is more effective than other methods.

In 3D worlds, which are truly immersive, this is all particularly true, because you can use your spatial memory in these environments. This type of learning is also connected to the effectiveness of a memorisation technique called 'memory palace'. You might have heard of this in relation to poker players, who use this technique to memorise the order in which cards are dealt, or in relation to mentalists, who use a memory palace to help them remember incredibly long prime numbers and sequences. Essentially a memory palace is a technique that involves connecting a piece of two-dimensional information, such as, for example, 'the queen of spades' in a deck of poker cards, to an imaginary journey through a specific spatial place – let's say, 'passing on a bridge' – and mentally remembering the imaginary journey constructed by all the imaginary places connected to all the cards drawn from the deck rather than the cards themselves. The memory palace technique is so effective in helping someone memorise complex information or sequences such as counting cards because, as *Homo sapiens*, our brain evolved through the centuries from primates whose survival was linked to remembering the location of the closest source of food, the closest shelter, the closest source of water and so on. Our brain biologically evolved to store memories linked to spatial information and actions in order to guarantee the highest chances of survival. Hence, spatial memory and learning by doing are incredibly powerful tools for conveying information and to guarantee memory retention in the most natural and effective way possible.

Why is all of this so important? Because the value of society is directly proportional to the level of scholarship that its population has. Therefore, being able to convey information in a more memorable way could help educate more people to a higher level and therefore help societies evolve.

Immersive learning isn't only about learning facts and information; it also allows us to improve our empathy. This is particularly true of virtual reality (VR) and augmented reality (AR) technologies, which are fundamentally empathy machines and are already being used for this purpose. For example, the United Nations is using VR to spread awareness of social issues such as the tragedy of the Syrian war refugees. With this technology, it is now possible for people to spend a day in the life of someone from a different group, such as for a man to experience what it's like for a woman to be in a male-dominated environment where she is potentially harassed or treated differently.

There are many benefits that can come from using this technology to enhance our learning. Of course, there are also negatives if the technology is misused, and we will explore some of these as we move through the rest of this book.

Ultimately, this is technology that will blur the lines between fun and work in the future. Game mechanics have the potential to transform boring or difficult tasks, as well as learning, into something more engaging and fun. In doing so, game mechanics have the potential to create better lifestyles for all of us and make us happier personally and professionally.

Gaming: offering a glimpse of the future for the Metaverse

Developments in the world of gaming have long paved the way for technological developments that become more widely adopted throughout society. This means the gaming world is a good place to look if you want to have a glimpse of what the Metaverse of the future could look like.

With specific reference to graphics, we are already seeing more photorealism, not only for avatars, but also environments, illumination and rendering. This trend is unstoppable and will only continue. You only have to look at the graphics in some of the games being released as I write this, such as *Gran Turismo 7* and *Horizon Forbidden West*, to see how far we have come in the last two decades when it comes to graphics. The graphics in these games are absolutely incredible.

We're seeing a drastic shift in graphic fidelity in tools such as Unreal Engine 5, and this will not only be the kind of graphic fidelity we will see in future videogames, but also in the Metaverse. However, it is important to make the distinction that the Metaverse is for the masses. The level of computing power we are currently using for modern consoles and gaming PCs is not what most of the population has access to, and gamers are just a tiny fraction of the people the Metaverse will reach. This highlights the importance for more computing to be carried out in the cloud, rather than on local devices, because that will give more people access to a high level of graphic fidelity within the Metaverse.

I envisage a future where, instead of buying expensive devices that require a lot of components to deliver this computing power, users will be able to buy inexpensive and smaller devices and then pay for a subscription that will give them access to the computing power 'as a service' through the cloud. When we reach this point, we will see better graphic fidelity for everyone in the Metaverse. However, we are not there yet, and it is likely to take at least 5–10 years before we reach this point.

Obviously, this is important if you want your Metaverse platform to be successful – because, the lower the specs, the more people can access it. Of course, this doesn't only apply to graphics, because there are multiple factors that require computing power, such as physics, AI-driven nonplaying characters, multiple concurrent users, persistency of information, etc. Ultimately, however, the world of gaming is giving us a glimpse into how the Metaverse is likely to visually look in the future, with improved graphic fidelity and photorealism. It's no exaggeration to say that the aim is to create something akin to Star Trek's Holodeck – just imagine the experiences this will present! This is where the Metaverse is heading; the future is in sight, if still a little way out of reach.

CHAPTER 5
AN UNHACKABLE AND DECENTRALISED REVOLUTION – ENTER BLOCKCHAIN AND CRYPTO

With contributions from Jamie Solomon

Jamie Solomon is an innovator with over 40 years of experience in implementing the 'art of the possible' to deliver business value in global consulting firms. He has consistently leveraged cutting-edge technology to drive transformation both within individual companies and at an industry level. He is responsible for Blockchain and Multi-Party Systems for Growth Markets and runs Accenture's global NFT and Token Economies business. He has led programmes in the blockchain space for seven years. Jamie led the creation of several financial services consortia (EquiLend, Clarient, RiskStream Collaborative) that fundamentally changed the way the industry works. Jamie and his wife Angelina, a high-end fashion designer, have three absolutely adorable kids (all with four furry feet) – two Siberian Huskies and one Bengal cat. Jamie enjoys sports of all sorts, including playing and watching basketball, soccer, football and boxing.

To fully understand the Metaverse and to begin to comprehend many of its use cases that have so far been identified, you need to understand the concept of blockchain. A blockchain is an integral part of the fabric of the Metaverse. More specifically, it provides the foundations for the economic framework of the Metaverse.

What is blockchain, and how does it work?

Blockchain is a distributed database that is shared amongst the nodes of a network. However, blockchain functions differently to a traditional database. A regular database stores the data as rows and columns, whereas data is stored as blocks in blockchain.

To picture how blockchain works, imagine that you and five friends are going to share your CDs with one another (I know, we're going back to the 1990s!). Each of you has a cog (this is the network node) that represents you, and all of these cogs are connected by a chain (the network). When you share a CD with one friend, your cog moves and their cog moves, but so do the other three people's cogs in your system. Each movement creates a record of which CD was lent to whom – let's imagine that this record is represented by a ring that is added to the bottom of each gear shaft.

This means that each person in this system, which is connected by a single chain, has the same rings on the gear shafts under their cogs and therefore they all have the same information. So, each node in a blockchain system is like one of these cogs with its rings, and the system is connected by a single chain.

The fundamental difference between a blockchain and a database, whether it's a standard centralised database or a distributed database, is that, with blockchain, the data that each node holds is guaranteed to be the same. This is the case whether the data is part of a private or a public network. This creates an unprecedented level of trust in that data.

This is also why blockchain is extremely difficult to hack, because not only is the information in each node encrypted, but this information is also duplicated across every node in the database, and they are all connected. In addition, each additional 'ring' of data is encrypted based on the previous one in the chain. This means if you want to hack transaction 253, for example, you not only need to hack the other 252 blocks in that chain on that one node, but you need to do this across every node in that database. This is why blockchain technology is so great when it comes to safety and security.

The difference between public and private blockchains

In a public blockchain, everyone can see everything, which means all of the parties who are part of this network see the same information. In a private blockchain, meanwhile, only those who are signatories to a specific block will be able to see what it contains. Bitcoin and Ethereum are examples of public blockchains, while Corda or Hyperledger Fabric are examples of private blockchains. What differs between the public and private blockchains is not the information itself, but who can access that information.

There are, naturally, some situations when private blockchains are preferable. Those national stock exchanges/clearing/ settlement agencies that are being built on blockchain are built on private-permission blockchains, because those transactions are legally required to be private. For example, Goldman Sachs should not be able to see the trades being made by Morgan Stanley, and vice versa, which is why private blockchains are used in this sector. These requirements don't exist for a public blockchain like Bitcoin or Ethereum, which is why they are open for all to see.

Where does blockchain come from?

The concept of blockchain originated in the mid-1990s through work carried out in the academic community, but it didn't come to widespread attention or prominence until the publication of Satoshi Nakamoto's whitepaper in 2008. In this paper, the anonymous Satoshi outlined the creation of a new online payment system, the world's first cryptocurrency: bitcoin.

The vision was for a currency that wouldn't be limited by national boundaries or the regulations in a particular jurisdiction. It would be global, available for everyone and provide equal opportunity for all. Blockchain was developed alongside bitcoin to create the trust required for the cryptocurrency to work. Being trustworthy is the most important function of blockchain. It not only means that everybody sees the same information, but it also ensures that everybody knows that the information they see is valid and correct.

You may have seen stories about blockchain being hacked and about people suffering losses in the blockchain space (particularly the public blockchain space, as you don't see the same issues in the private blockchain space). The reason these issues occur in the public blockchain space is not due to weaknesses in the blockchain technology, but because of social engineering and people giving away their private keys, which in turn gives someone else access to their data.

In fact, all of the hacks that occur do not happen in the blockchain itself, but through the mobile apps and websites that connect with a blockchain server to push the transactions executed by the users. The weak point is not in the blockchain, but in the connection between the mobile phone/terminals and the server. The blockchain itself is unhackable; however, the infrastructure that supports transactions in blockchain is very much hackable – this is where the issues lie and where technical architects still need to do some work to improve security.

Blockchain use case: cryptocurrency

If we take bitcoin as an example, we can see why blockchain is so important for cryptocurrency. At present, there are 21 million bitcoins in the world. There will never be more bitcoins than this, because, unlike the US dollar, no more can be created. The Federal Reserve in the United States can (and does) decide to print more dollars and put them into circulation. This can't happen with bitcoin.

If you want to pay for something with bitcoin, which is obviously digital in its form, you need to be able to demonstrate that, when you give someone bitcoins from your e-wallet, they aren't also going to be duplicated in anyone else's e-wallet. Blockchain is the technology that ensures bitcoins can't be double-spent.

Blockchain use case: NFTs

Non-fungible tokens (NFTs) are Immutable certificates of ownership for any digital assets. Blockchain ensures that, just as with the bitcoin example, it is possible to know who owns a specific digital asset. 'Non-fungible', by definition, means something is unique, and therefore there can be no other copy of a given NFT in existence in the world. NFTs, along with their subset digital twins, are superb uses for blockchain.

There are multiple ways in which NFTs and digital twins have the ability to change the way the world works. For example, you could buy a car in the real world and also receive a digital twin. The digital twin is a special kind of NFT that can communicate with a physical asset. It contains a series of smart contracts that enable it to do all kinds of things. A digital twin of your car could help you find parking spaces and pay for them automatically; it could allow you to purchase additional horsepower for your car; it could automatically book the car in for its annual service; or, if the car breaks down or is damaged in an accident, it could preorder the parts required to fix it and have those sent to the repair shop while the car is being transported there.

Essentially, when you change the digital version of the asset, it will modify the physical version and vice versa, which is why the NFT can allow you to do things like increase or decrease the horsepower of your car, for example. When it comes to the auto market, cars are governed by a standard that's promulgated by the Mobility Open Blockchain Initiative (MOBI). Within the auto industry, 158 organisations subscribe to MOBI, which includes almost all vehicle manufacturers. The blockchain standard set out by MOBI governs how those NFTs communicate with their particular vehicles.

Within the auto industry, NFTs can have further applications. For example, if you purchase a car with an associated NFT that is connected to the car's ignition, then you would only be able to start the car if you had the NFT. The NFT would sit on an application built by the automaker, so when your phone connects to the car via Bluetooth and it recognises the NFT on your mobile phone, you can start the car.

An NFT can also contain the full history of a vehicle, because you can enrich the data on the original NFT with additional NFTs. For example, if a sensor in your car perceived that you'd had an accident, that you had replaced a part or even that there is a defect with a part, the underlying technology allows that information to be attached to the NFT, giving any future owner a fully comprehensive history of the vehicle, which means they know exactly what they are buying. With NFTs, automakers could also receive a percentage of any resales of their vehicles. This has huge financial implications, as there is a massive amount of untapped revenue that can be made available to automakers, but this also benefits the buyer of the car, who will know exactly how many kilometres the car has been driven, what parts have been changed, whether the vehicle has been involved in any accidents and so on.

In this industry, NFTs could almost eliminate auto theft and would seriously limit chop shops, because every NFT is a composite of all the physical components of a car. This means not only would the introduction of NFTs greatly reduce people's ability to steal cars, because you can't drive them, but it would also mean the thieves couldn't reuse any of the parts that are connected to the NFT (which would be any part with an electronic component).

NFTs and digital twins are also appealing for luxury goods manufacturers. Let's say you buy a Dior handbag and that you also receive an NFT with that bag. As a consumer, this proves your ownership of that specific Dior handbag, and that means, if you sell it, you can prove the authenticity of the bag.

For the manufacturer (in this case, Dior), there are several beneficial use cases. The first is that NFTs that show the authenticity of their handbags will eliminate the huge market for knock-off designer bags. More importantly, the NFT also provides Dior with a secondary revenue stream because, whenever the bag is sold, Dior can receive a percentage of the sale price. So, a luxury goods manufacturer cannot only reduce the damage caused by fraud and fake products, but also generate a new revenue stream because, at present, they aren't able to participate in the resale market. For handbags alone, that resale market is worth $70 billion a year, so it presents a significant opportunity.

Another benefit for you as the consumer is that the NFT can potentially give you the right to take a virtual copy of your Dior bag into any of the virtual worlds you participate in. That means you can use the bag as part of your avatar and lifestyle within *Decentraland, Horizon Worlds, The Sandbox* or any virtual world you're part of.

> *The Sandbox: A community-driven platform where voxel assets and gaming experiences can be created and monetised by the platform's users. SAND is the crypto token used within the platform, and LANDS are the parts of the world that are owned by users to create and monetise experiences.*

This is how NFTs connect to the Metaverse, because it will allow us to combine our physical and digital worlds, bringing what we love in real life into our digital lives as well.

Rights associated with NFTs

When any NFT is created, there are four categories of rights that can be conferred by purchasing that NFT. The rights you're afforded will vary depending on the NFT in question.

IP rights: An NFT might only give you the right to view and display an item; it could confer the right to use the NFT commercially; or (as in the case of the Dior bag) it might allow you to own the underlying asset, or even modify it.

Financial rights: You may have the right to sell an NFT; you could have the right to loan or borrow against a particular NFT; or you might have the right to royalties. US musician Nas has sold the royalty rights to two of his songs as NFTs, which will allow those who purchase the NFTs to receive a percentage of streaming royalties for the associated songs[1].

Access rights: An NFT could give you the right to enter a virtual world, play a game or attend a special event.

Redemption rights: This could give you the right to a physical good (as in the Dior case), or purchasing a particular NFT could give you the right to loyalty rewards.

These rights are all enabled by smart contracts, which are written on top of the blockchain. This is why the Metaverse and blockchain are inextricably tied together.

NFTs are an important step in the evolution of the Internet of Things (IoT), because, when all of the components of a physical item are connected via a single NFT, you create a digital footprint of the physical asset. We talked earlier about how NFTs could revolutionise the auto industry, but just think of the real estate market too. A house is an extremely valuable asset, and, often, when you sell a property on, the value has increased. Construction companies currently don't profit from the resale market, but with NFTs they could. This technology is going to create entirely new markets for many traditional industries.

The advent of the creator economy

Many people are describing the Metaverse and NFTs as the advent of the creator economy, because now there is technology available to allow creators to benefit from their creations beyond a single sale. Creators can range from manufacturers (like automakers) to artists, musicians and fashion designers.

In this new world, the original creator of any item would potentially be able to receive secondary and tertiary royalty rates from future sales of their creations.

This also connects to the movement towards Web3, which is a new evolution of the web as defined by Ethereum's co-founder Gavin Wood in 2014. His vision is essentially the full decentralisation of the Internet, whereby we move towards an Internet that isn't based on a centralised database but instead where every database is decentralised. The beauty of this approach is not only in the use cases we've already discussed with cryptocurrencies and NFTs, but also in terms of coding.

So, the code on the Internet and certain platforms can be connected to blockchain systems. This means that, as a coder, you can attach a piece of your code to a specific block and not only continue using your code yourself, but also allow others to build on top of your code. When someone else builds on your code and makes money from it, you could receive some money as well. It's a way of fairly distributing royalties among everyone who has contributed to a specific piece of work.

All of this is hugely important for the creative economy, because it means that anyone who is a creator stands to benefit from whatever they create long after any initial sale or work is completed. In essence, it's a model for funding innovation and creation.

The economy of the Metaverse

Blockchain and the Metaverse are entirely separate concepts, yet certain aspects and specific use cases of the economy of the Metaverse will not function without blockchain technology at its core on an economic level. What blockchain provides is trust, which can be connected to identity, ownership and the correct distribution of meritocratic wealth in the form of royalties and ownership rights. It creates order in the chaos.

There are three ways in which the economy usually works in Metaverse platforms. The first is a bottom-up approach, where some of the users on the Metaverse platform (known as *creators*) design, create and sell digital assets within that Metaverse platform in the form of a smart contract that certifies the ownership of that specific unique digital item, also known as a *non-fungible token* (NFT). The other inhabitants of this Metaverse can then purchase those assets and make the economy circulate in a self-sustaining way.

The second way in which the economy works in the Metaverse is a top-down approach, where brands join Metaverse platforms by either renting the land from existing landowners or buying their own land to increase their brand awareness, eventually becoming creators and dropping their branded NFTs into the community from the top and making money from them in a top-down model. Some brands will enter the

Metaverse simply to raise their brand awareness rather than to sell any assets or goods.

The third way that we see the economy of the Metaverse operating is also, we believe, the most interesting. This is a top-down and bottom-up economy. In this model, brands will drop NFTs into a Metaverse platform that are linked to the rights of using their intellectual property (IP) rights, which can be purchased or, in some cases, acquired for free by the creators on that Metaverse platform. Once the creators have these IP rights, they can build upon that NFT and create new assets that can be sold on the Metaverse platform. Of course, there will be terms and conditions attached to those IP rights, but it's ultimately an opportunity for brands to allow creators in the Metaverse to become the designers of their products' future in a digital realm and exponentially amplify the products and services linked to their brand.

Coca-Cola is one of the big brands that is already experimenting with this approach. The company dropped the rights to use its logo, and creators have already created all manner of assets and merchandise – from Coca-Cola t-shirts to a Coca-Cola club – all produced as NFTs.

The beauty of the Metaverse is that brands can decide how they want to participate and whether they would prefer to stick to a top-down approach across the board, or whether they're happy to relinquish control of certain brand elements and give creators more freedom over how they use their brand in the Metaverse. This is where blockchain again becomes essential, because the smart contracts that underpin the branded NFTs allow the brands to protect themselves as they're revocable. Revoking the rights from the blockchain will lead to the revoking of all the assets linked to any specific revoked NFT. All of these elements are created and stored on the blockchain.

Another interesting and unique concept linked to the Metaverse and enabled by blockchain technology is what is called a *decentralised autonomous organisation* (DAO). The Metaverse platform *Decentraland* is probably the most famous example of DAOs, and its peculiarity is due to the way in which it is governed. In a centralised (non-blockchain-based) platform like Meta's *Horizon Worlds*, governance is regulated by a leadership management team and, in the case of a publicly listed company, a board of investors. In *Decentraland*, all the decisions about its governance and business are voted on in a direct democracy fashion by the users who have purchased a portion of digital land on the Metaverse platform as NFTs and have voting rights proportional to the amount of land owned. As a result, it is possible to steer the direction of the platform financially, technologically and even politically, because the users have that power through their votes. This is revolutionary, because it hands power to the people and allows them to decide on the platform's future.

Blockchain technology is at the heart of this revolutionary approach, because it underpins the smart contracts that allow rights to be passed between those interacting in this world. Blockchain provides the trust in the process and the protection that you need as both seller and consumer, because, whenever you buy anything in that world, you know exactly what you're getting, you know it's authentic and you know you're the only one who is purchasing it. Blockchain is fundamental for providing this level of trust in transactions in the Metaverse.

Identifying the risks

For all that this is good, there are also concerns about this model, particularly around areas like governance and service-level agreements. It isn't appropriate for every organisation that engages with the Metaverse to be a DAO. Financial organisations are a good example, because they are required under their fiduciary responsibilities to have consistent governance. The DAO model isn't appropriate in this scenario – just imagine what might happen if the New York Stock Exchange became a DAO and, all of a sudden, the DAO changed the rules. That would be dangerous – not only for those directly involved, but also for the economy.

While the Metaverse will allow us to do many wonderful things, this has to be balanced with an understanding of how to develop and use it in a safe and responsible way that protects the security and rights of the participants.

Other risks that we can identify within the Metaverse at the time of this writing relate to NFTs and cryptocurrency. Let's start with NFTs. Like any investment asset, NFTs are only worth what someone is willing to pay for them. We have already seen examples where an NFT is sold for millions, only for the buyer to try to resell it at a later date and discover that its value has dropped substantially, because interest has dropped off completely – Jack Dorsey's first tweet being a classic example. It was auctioned for $2.9 million in 2021, but, when the buyer tried to resell it in April 2022, he was only offered just over $6,200[2].

Many NFT initiatives we are seeing at the time of this writing are being driven by massive social media campaigns, celebrity endorsements and influencers. This can inflate the value of the assets that are being sold, and, naturally, when the wave of interest dies down, so too does the value of the NFT in question.

We have also seen a massive crash of certain cryptocurrencies in 2022 (Terra Luna being one of the most high-profile examples). While none of the cryptocurrencies affected at the time of this writing are specifically associated with any existing Metaverse platform, that's not to say this isn't a risk to be aware of in the future. The value of these coins is based on supply and demand, and their value can fluctuate massively (as we've seen). It's therefore important for you to understand that, when you convert your hard-earned money into a cryptocurrency that is accepted on a Metaverse platform, you are making an investment. Investments can go well, or they can go badly, and this is a risk you take when you start using digital currencies.

When we enter the Metaverse, we also have to be alert for scams. Most of us are used to deleting the emails from a Nigerian prince that still land in our inboxes from time to time, promising us a share of his millions if you only give him your bank details! Similar kinds of scams can happen on Metaverse platforms, but in a different guise. You might be promised digital items, loot or other wonderful things by someone in the

Metaverse in exchange for money, but they suddenly disappear when you give them your money (and you don't get the items you were promised).

These scams are quite common in videogames that use game currencies, such as *World of Warcraft* and *Diablo*, but they can happen in the Metaverse as well, and we all need to be alert to them as we start to spend more time on these virtual platforms.

The key is to be aware of these risks and to treat putting money into cryptocurrencies, NFTs or other digital assets as similar to any investment you might make in the real world. Do your research, know what you are investing in and apply some common sense. As I write this, a lot of these platforms are unregulated, which presents opportunities for con artists and scammers, but these are risks we can mitigate by making sure we are informed about our investments.

We also have to understand that there are a number of very real risks associated with the Metaverse (as with any emerging technology), including those that we can't currently foresee, such as in relation to governance, privacy, moderation and misuse of the technology. Being aware of the potential for blind spots will help us detect these risks sooner. Identifying these risks as early as possible is essential, because knowing about them will help us to find solutions. However, we also need to accept that there will be some risks that we are completely unaware of at the moment, and we need to be vigilant to spot them as the technology evolves.

Blockchain: creating freedom in the Metaverse

There are several advantages to utilising blockchain – in fact, blockchain is key for the economy and market structures within the Metaverse. It also enables the creation of DAOs, which gives power to the people and therefore power for users on certain Metaverse platforms to steer their direction.

Blockchain technology can also be used to determine digital identity, which is essential for creating accountability within the Metaverse and for ensuring that everyone in any given Metaverse platform abides by its rules. Setting rules within the Metaverse is a huge topic, and one we'll return to in more detail in Chapter 7.

Metaverse and NFTs: revolutionising how brands operate

The Metaverse, and NFTs in particular give brands the ability to invert their current go-to market structure. For example, sports teams like Liverpool Football Club sell their merchandise through a host of third parties, they sell tickets through third parties and even interact with fans through third parties. This is simply how the world works at present.

Tomorrow, with the arrival of the Metaverse and this new technology, Liverpool (and any other brand) will be in a position to take full control of their assets and determine how they want to market, sell and direct those towards their customers. They can build a customer engagement model that allows them to engage *directly* with their customers, sharing the message that they want to deliver and retaining full control over that. How is this possible? Through smart contracts and NFTs. As a brand, you will suddenly be in total control over who has access to your brand assets, as well as what messages and rights they receive alongside those assets.

In this new world, a company can create its own brand strategy and completely own it. Companies will no longer need to rely on third parties to distribute their goods, because they will have the means to control all of that themselves.

Major brands around the world have already recognised how revolutionary this is. They are asking how they can take advantage of this and how they can build brand loyalty and change their customers' actions. The ability to change the customer interaction model through the use of NFTs and the Metaverse is completely game changing, whether you are a luxury goods manufacturer, a bank or a small business.

Opening up a global marketplace

One of the most exciting aspects of the Metaverse is that, for the first time in the history of humanity, the opportunity to build a business; create, market and sell products globally; and, in the process, potentially make a good income (or, in some cases, a fortune) is completely independent of someone's location or postcode of birth.

Wherever you live in the world – whether that's Namibia, Uganda, the United Kingdom, Thailand or Zimbabwe – if you have an idea for a digital product that works and that people will pay for, you can make a lot of money. This is a great leveller, because geography doesn't matter in the Metaverse and, once we start using cryptocurrencies more widely, neither will the strength of your home country's currency. This is how the Metaverse is shifting us towards a creator's economy, because content is what people want in the Metaverse.

You don't need to be smart to create an NFT, but what you do need is to be creative, to be able to come up with digital products that other people want and to run effective campaigns across social platforms like Tik Tok, Instagram, Discord, Reddit and Twitch to market them through communities. This is great for creators and small businesses, who will have access to a global audience and the ability to produce and sell their products at a very low cost.

This model not only gives power to the creator, but also strips away a lot of the unnecessary things that we all have to pay for in today's economy, because it removes intermediaries from the equation. With blockchain, you'll no longer need an organisation to research the deeds on a property, for example, because all of that data will be stored in one place and be readily accessible. If you're a business that works as an aggregator or intermediary, you will likely need to pivot and change your business model to survive in this new economy within the Metaverse. The key is to ensure that what you offer are value-adding activities. Any activities that don't add value to people's lives or the service they're buying will disappear.

For industries that haven't changed the way they operate for centuries, such as banking, this is a scary prospect. Business as usual is going to come to an end, and the organisations in this sector are going to need to adapt to survive. Organisations in the banking industry are only just beginning to see the threat, and those that are forward-thinking are already exploring what their role can be in the future with the arrival of the Metaverse, blockchain and cryptocurrencies.

Some are even dipping their toes into the waters of the Metaverse – for example, by creating a spinoff of their company with blockchain. Bitcoin exchange companies, meanwhile, are working on providing services for NFT purchases and land brokerage in the Metaverse. But these are the digital primitive use cases. Just imagine what kinds of services and products we'll be able to create in just 10 years from now, when we really understand the value of decentralised technology and the social mechanics that it enables.

What could become the Facebook, Tinder or Uber of the Metaverse that's connected to this decentralised economy? Right now, nobody knows, but imagine what this technology and these businesses have the potential to be. The possibilities have yet to be defined.

However, what we do know is that blockchain provides security for transactions conducted within the Metaverse, and, based on blockchain, we have a structure for how commerce will work in the Metaverse. Where it develops from here and how that shapes the future of the Metaverse is, right now, anyone's guess. We are entering a brave new world.

Endnotes

1. Shutler A (2022) 'Nas to sell royalty rights to two of his songs as NFTs', *NME*, 9 January, available at: https://www.nme.com/news/music/nas-to-sell-royalty-rights-to-two-of-his-songs-as-nfts-3133351
2. Liang A (2022) 'NFT of Jack Dorsey's first tweet struggles to sell', *BBC News*, 14 April, available at: https://www.bbc.co.uk/news/business-61102759

CHAPTER 6
A BRAVE NEW WORLD

There are different schools of thought about the Metaverse and what makes it the Metaverse. There are some people who believe the Metaverse needs to be social in nature, and that it would not exist without interaction from multiple people. There are others who believe that the Metaverse needs to have some kind of transactional commerce base behind it, such as blockchain, crypto or non-fungible tokens (NFTs). These are not mutually exclusive concepts though, and I believe both will be fundamental to creating the Metaverse of the future as they will fuel each other.

We also need to understand that there are technologies that support the Metaverse and are foundational to its purpose, such as immersive worlds and the potential for connection to augmented reality (AR) and virtual reality (VR) technologies, and these can have an incredibly positive impact on the value chain of any company anywhere in the world. In fact, there are a multitude of enterprise use cases for the Metaverse already, many of which do not focus on business-to-consumer inter-actions and instead are centred on operational efficiency and productivity.

To help you understand how this brave new world we're entering could look, I'll explore some of the use cases we have already seen emerging in the Metaverse and show you just how vast the possibilities that lie before us are.

Product design and development

The majority of companies that are designing and producing products for consumers do so using computer-aided design

(CAD) to create 3D models. I'm talking about everything from shoes to automobiles. The design process is often collaborative in nature, and, once a 3D model has been created, there will be a collaborative review of those design files. Working with these kinds of 3D models can sometimes be difficult. In some cases, businesses may need to print clay prototypes that are not only incredibly expensive, but also difficult to make changes to in real time.

This also means that teams of people who are not all in the same place can find it difficult, if not impossible, to work on the same model at the same time, while communicating and discussing potential issues and their solutions. Enter the Metaverse.

The technologies that are powering the Metaverse, such as multi-user experiences, 3D worlds and immersive technologies, can enable collaborative design and review for non-co-located teams. This essentially creates a virtual workspace, but one where everyone can see and work on the same 3D model at the same time, while communicating openly.

Another advantage to using virtual workspaces and 3D models for product development and design is that it is far easier to share those 3D models with focus groups or consumers to allow them to review the product in real time and provide feedback quickly and efficiently. If you have permission from those who form your focus groups, you can even monitor physiological responses to products by tracking gaze position, body posture and tone of voice. This allows you, as a business, to track the emotional responses of your users.

This opens up a wealth of possible use cases just related to product design and development:

- Real-time product review
- Collaboration on product design among non-co-located teams
- Carrying out user assessment and acceptance testing using digital models without the need to produce physical prototypes.

When it comes to user testing, this could be invaluable, giving your business deep insights into how to improve the ergonomics of its product, its usability or even just the product's aesthetics.

Manufacturing: training

Once a 3D model has been produced and agreed upon, the product itself will need to be manufactured. In order to create a new product, any company will need to level up the knowledge of its workforce. In any manufacturing company, employees will need to complete training before they are able to start producing a new product. This ensures that they understand the components required in what can be complex devices, such as smartphones, TVs or cars. Of course, there are elements of this production process that are completely automated and carried out by robots, but other steps need to be completed by hand. To have an effective and efficient production line, your workforce needs to be upskilled to be proficient with the new production methodologies.

There are two ways of doing this. The first is the old way of training people by bringing them to a location with a dummy production line where they can be taught how to assemble the products. For this to work, you not only need a large physical space but also for that space to be easily accessible for everyone you want to train. In addition to the training itself, you also need people to monitor the trainees' progress and performance. How scalable is this approach? Not particularly, especially during a global pandemic when people's mobility and presence at work is affected to the point that you cannot have many people together in a single space simultaneously. This is where the immersive technologies of the Metaverse come in.

Companies can use the same kind of 3D model that they are already using for the design and design review processes, as well as for user experience testing, to create training modules for their staff. In doing so, they can upskill their employees to enable them to physically make the product. Of course, delivering and distributing training is made much easier when it is in a digital format.

Imagine that you could send an email or push notification to the mobile phone of every employee who needs to complete the training, telling them to put on their headset, log into their corporate account and complete the training in an immersive space. Studies conducted at Stanford University through the immersive learning platform provider Strivr have found that those completing training using its VR solution reported up to 15% higher memory retention rate when compared to normal e-learning methods delivered via 2D screens. This demonstrates that learning by doing is important to improve the stickiness of the information being shared. In addition, the VR training solution led to 30 per cent higher employee satisfaction, and those who undertook the training scored higher on tests 70 per cent of the time[1].

For those undertaking the training in a 3D immersive environment, as opposed to with 2D training materials, the experience will better prepare them for stressful situations. It also allows them to develop muscle memory without requiring the company to buy and maintain additional equipment that is simply used for training purposes.

At the time of writing, YouTube is the largest repository of 'how-to' videos in the world, and is estimated that approximately half of all searches on YouTube are for 'how-to' videos. We already know that people are seeking out visual content because they find it easier to learn in this way than from reading manuals. Now imagine that you could combine the benefits of visual learning with simulating actual tasks to help people develop muscle memory, understand how to use their tools in real time and create the same kind of emotional engagement that is experienced when someone is working in the real world.

As if all of these benefits were not enough, approaching training in this way and making full use of virtual tools and training environments allow your business to access analytics data to track each individual employee's performance. This will immediately give you a clear view of the proficiency of your workforce, thereby allowing you to more effectively plan your production cycles.

For example, if you need to have 70 per cent proficiency among your workforce in order for it to be viable to set up a new production line, you will be able to track how your workforce is improving as they move through the training. This will enable you to produce a fairly accurate timeline for when you can start producing your new product. Training in a virtual, immersive environment also allows you to troubleshoot before people are working on physical products. Through the use of this technology, you can identify where people tend to make errors in the process and refine it, to ensure those errors are not repeated when you go into production in the real world.

There are significant cost savings to be made by taking this approach to your training, not only because you will be saving money by not needing a physical location in which to conduct the training or to pay for people to travel to that location, but also by not having to create a physical training environment. You will also save money on the trainer, and your training will instantly become more scalable, because you can distribute it to your workforce via devices that are readily available now. To deliver training in this way, you need a device that costs about one-fourth the price of a top-of-the-line iPhone available in 2022.

Although I've talked about physical training in this section, immersive 3D training environments are also incredibly valuable for emotional intelligence training. Not only does this kind of training make it more enjoyable for those participating, but it also gives them a distraction-free environment, making the whole process more effective.

Manufacturing: production

Once you have trained your workforce in the manufacturing of a new product, you can expect them to be around 75 per cent proficient in the task, as compared to someone who has been making these products for around one year. How do you bridge that 25 per cent gap between someone who is fully proficient, which takes time and repetition, and someone starting out with a good grasp but not full proficiency? The answer is, by using the Metaverse.

Naturally, businesses want this 25 per cent gap to be closed as quickly as possible to ensure that everyone on their manufacturing team is working at full efficiency. This is where you can use just-in-time training. This requires each person to wear a head-mounted device, like a Microsoft Hololens 2, that provides each individual with step-by-step instructions about how to complete a task using the same 3D assets that they've already used in their business for design, review, user experience and training.

The huge advantage to just-in-time training delivered by a head-mounted device is that it means the individual following the training doesn't have to keep switching their attention from the work plane to the instruction plane, because everything is driven from the glasses that are literally in front of their eyes. Another benefit to this approach is that the amount of information a user can consume and process is considerably higher when it is delivered visually rather than by reading a manual.

This also makes it quicker for someone to complete a task, particularly if it is complex. Just think about how much time you spend going back and forth between instructions and being in the work plane when you are completing a new task. Now imagine that the task you are working on involves wiring the fuse box of a car, where you have multiple cables that all have to be in precisely the right place.

Not only can this kind of system make it quicker and more efficient for someone to complete the task, but it can also make it safer. Using the camera in the headset, the system itself can perform safety and quality checks while the person is working. In time, it will be possible to enable advanced analytics capabilities that can predict human errors, recognise when someone is about to make a mistake and correct them before they either put themselves in danger or damage the quality of the product. You will make the work environment safer and improve the quality of your product at the same time.

Branding and marketing

Once your product is built, you'll need to market it, and once again we can return to the 3D model of that product, as well as to the Metaverse.

Imagine taking that 3D model and dropping it into a Metaverse platform. You can then use it to create interesting and entertaining experiences that not only showcase the new product but also your brand to great effect. We have to remember that the Metaverse is centred on experiences and delivering emotional value to the users who are in that specific part of the Metaverse.

It is up to companies to be creative enough to promote their products in the Metaverse, raise brand and product awareness and publicise any new products in such a way that they are appealing and palatable to consumers. This is a topic we will cover in much greater detail in Chapter 8.

All of the use cases I have just explored relate to designing, producing and selling physical products, but what about the potential for creating full digital products in the Metaverse?

Digital products

This is a use case of its own and one that businesses are showing an increasing interest in. As we discussed in Chapter 5, companies have the ability to create products that are fully digital and only exist in the Metaverse. For example, a company could create a pair of sneakers that is fully branded and that only exists in the Metaverse, with no real-world equivalent. We have already seen many companies experimenting with digital products, particularly in the luxury space.

The rise in the global sale of NFTs and digital goods, with more than $20 billion worth of digital items traded just in Q3 2021, clearly demonstrate that that consumers are happy to shell out money to receive exclusive digital content that is theirs and theirs alone. This is all possible thanks to the exclusivity ignited by NFT technology. Whether it be limited-edition digital sneakers or limited-edition digital bags, people are already spending thousands of dollars on these kinds of products. This is a particularly valid use case for any business that already has a strong brand that appeals to the largest audience you can find in the Metaverse as I write this book – namely, those who are part of Gen Y (also known as 'millennials', born between 1981 and 1996, and grew up as the Internet evolved) and Gen Z (those born in the late 1990s and early 2000s, who have never known a world without the Internet). Releasing digital products not only capitalises on existing brand awareness, but also enables these brands to gain even more followers due to the fact that the people buying these digital products become brand ambassadors.

As I explained in the previous chapter, the economy within the Metaverse works in two ways. The first is top-down, where a brand drops digital products that users can buy directly from the brand. In addition to existing brands getting involved in this aspect of the Metaverse, there are also entirely digital fashion houses such as Fabricant and even digital fashion multi brand retailer such as DressX.

The second is bottom-up, where creators who also inhabit the Metaverse are building products themselves and selling them to other inhabitants of the Metaverse. If your products are cool and people want to buy them, you could build an entire business on this. Even if your products are sold and used within the Metaverse, as a business that doesn't mean you can't promote them on platforms outside of the Metaverse. You could use Instagram, TikTok, Twitch, Reddit, Telegram or Discord to promote your NFTs and digital products and encourage people to use them in the Metaverse.

Then there is the blend of these two approaches, where a brand collaborates with creators in the Metaverse, such as the example about Coca-Cola in the previous chapter.

Job hunting and recruitment

There are several use cases for the Metaverse within the category of recruitment, job hunting and HR generally. Through VR and 3D worlds created for enterprises, it will be possible to give someone a full 'day in the life of. . .' experience of a specific role within a business, for example. You could also use this technology to set a practical test for job applicants and assess their ability to perform in the role. For example, imagine that you are hiring someone for a role that is physically demanding. You hire someone and then, on their first day on the job, they realise that the role is too physically demanding for them. You have invested not only time in your recruitment process, but also money. However, this situation could be avoided entirely if you are able to offer them a virtual 'day in the life of' trial to assess whether they enjoy and are capable of doing the job.

Of course, you can also hold remote interviews and conduct vocational checks in VR in the Metaverse. It isn't only within recruitment that there are valuable use cases. If someone is coming out of university with, say, an engineering degree, they could have the experience of working in different roles within the engineering profession before deciding which path they would like to follow in their career.

VR can also be used for onboarding new employees. The consulting firm Accenture has embraced the Metaverse to onboard 120,000 employees during the pandemic. Accenture purchased 60,000 headsets from Meta and created a virtual onboarding experience using 3D worlds because they realised that carrying out onboarding in front of a wall of 200 people on a video call wasn't efficient. This is particularly true when it comes to creating the bubble of sociality that you usually have at onboarding events that are hosted in person and on location.

For people joining a consulting firm like Accenture, it is very important to start creating your network in the first few days as you are completing your induction, because the network you build will then give you access to your first assignment in the company. When you are faced with a wall of 200 people in a video call, it is very difficult to network. People simply resort to using the chat function, and that bubble of sociality is nearly impossible to recreate on video calls with so many people. However, the platform that Accenture has used to create its 3D world is very easy to use, and, because you have positional audio, you can create a small circle of people and talk very easily to one another. This kind of social fireplace gathering happens very naturally in this virtual 3D world, and it has been very successful. At the time of writing this book, Accenture has on-boarded around 150,000 people using this platform, which is remarkable in a company of around 700,000 people.

Creating a seamless digital journey

The Metaverse is often discussed in relation to consumer experiences, but it is important to recognise that the Metaverse should provide a seamless digital journey across every area of your life. This is a world where you can move from your word processor to gaming on your PC without any friction. There does not need to be any distinction between the 'work' element of the Metaverse and the 'entertainment' element of the Metaverse. We are talking about 3D and digital worlds, and balancing the different elements of the Metaverse is something that should come naturally to us.

In fact, we are already used to this kind of multitasking approach with technology. In real life, there are often situations where you have one screen displaying your emails, and another where you are watching a show on Netflix. You might also be chatting to your friends via your smartphone. Of course, there can be distractions, but this level of multitasking across different elements of our lives is something that comes naturally to us.

In the future, I envisage immersive technologies and devices, such as VR and AR glasses, allowing us to move seamlessly with our avatars from a work meeting to a game of golf with a friend, all in VR. Or you might go straight to a digital concert as soon as you finish your virtual call with a client or colleagues.

The workplace as the entry point to the Metaverse

We should remember that gaming entered people's homes through PCs that were purchased primarily for work purposes. I believe that the virtual workplace could very easily ignite a similar level of uptake in Metaverse technologies and, in fact, be the tipping point that pushes us towards the mass-market adoption of immersive technologies, such as the adoption of VR and AR glasses.

As I have already mentioned, Gen Z and Gen Y are the demographics most likely to be embracing the Metaverse and these immersive technologies at present, so businesses need to start thinking about how they can encourage older generations, such as Gen X and boomers, to start using this new technology. One of the best ways to encourage these demographics to make use of it is to show them there is a better way of working and collaborating in these digital realms, as well as to highlight the multitude of use cases connected to the certification of physical property and goods through digital certificates. In the previous chapter, we outlined many of the benefits of adopting blockchain technology and NFTs, and it is these benefits that we need to clearly communicate to all those who are unsure or hesitant about the Metaverse.

Blending the digital with the physical

One of the biggest and most important applications for the Metaverse will be blending the digital with the physical. I believe that the use cases that connect physical products to digital ones, the digital twins we discussed in the previous chapter, will be truly transformational for both the economy and for users.

When AR glasses become ubiquitous, we will have reached a point where the digital and physical really do merge seamlessly. Picture being able to see the digital world around you as you go about your day, with real-time information and analytics available to you. We will have superpowers that are currently unimaginable. Just think about how Google Maps has changed the way you navigate. If you have a mobile phone with an Internet connection, battery life and Google Maps, it is impossible to get lost. Even 20 years ago, having that kind of interactive and real-time navigational tool in your pocket was unimaginable. Smartphones have given us the ability to communicate wherever we are, with whoever we want.

The Metaverse is simply going to take all of these developments to the next level. Google Maps already allows us to move around in any street based on the 2D images in its database, but in the Metaverse you won't just be looking at images on a tiny screen and struggling to move the picture in the right direction using your fingers and a touchscreen. Instead, you could drop into a street anywhere in the world and look around you, experiencing it as though you were really there. Just imagine how amazing it would be to drop into a street in Florence and find yourself staring up at the Cathedral of Santa Maria del Fiore!

Or imagine a future where you are attending a networking event. Each person's name is visible just above their head (I forget names easily so I can't wait for this to be a reality!), and your glasses are feeding you information about who will be the most interesting people for you to talk to in that room. Your glasses could potentially even help you to interpret people's microexpressions to work out whether your conversation is going in the right direction.

Imagine the number of use cases where people would benefit from real-time microexpression analysis, from business meetings to dating, and even to help people who struggle to read other people's emotions to make more connections in social settings. This is a world of possibilities that will be enabled by information that is invisible to the external world, yet visible to you if you are wearing an immersive headset or similar device.

One of the reasons why this technology in particular has the potential to be so transformative is that it will not distract the person using it or create a barrier between them and the person they are interacting with. If you're having a conversation with someone in today's world, it is considered rude to pull your phone out of your pocket and start searching for information to fact-check what they are telling you, for example. However, with a wearable device, an application could run automatic fact checks on what you are being told and alert you in real time to anything that is false.

Tapping into new superpowers

I believe that mass adoption of the technology that will underpin the Metaverse and enable us to move seamlessly between the digital and physical worlds will only come once truly transformational use cases emerge and people begin to realise that, without an immersive device, they are missing out on superpowers that other people can access.

As I write this, we already have superpowers at our fingertips. If you want to hail a cab, you need Uber; if you want to go on a date, you need Tinder (although sadly you can't guarantee the quality of your date!); if you want to know what song is playing, you need Shazam. You need a mobile phone for all of those applications. With immersive devices like AR and VR glasses, these superpowers will not be at our fingertips but in front of our eyes. When we start to see use cases that you need these immersive devices to access, we will start to see much greater adoption of the technology. It will only be then that older generations truly start to see the value in this technology and also embrace it in their lives.

The Metaverse is not just a place where those who are part of younger generations can connect with brands and experiences. It will be an infrastructure that enables all of us to have superpowers. What is really exciting is that we don't yet know what those superpowers will be, because we are still very much in the digital primitive stage with this emerging technology.

You could also think of the Metaverse as a series of ingredients, which includes immersive technologies, blockchain, 3D worlds, community, identity, a new economy and so on. You can choose which flavours you mix together. You might blend all the flavours, or you may decide to carefully mix just two or three. It is like a magician's cookbook, and we are all still learning what we can whip up with these new and exotic ingredients. Some interesting recipes have already been uncovered, but we won't see any truly transformational recipes emerging until we experiment more with those ingredients. In turn, the ingredients themselves will ripen and become more flavoursome as the technology evolves to become more sophisticated and reliable. This will allow us to create entirely new recipes.

The bottom line is that you need to offer transformational experiences to push people to buy new hardware. This is something that Apple does incredibly well by ensuring that all of their product strategy is based around user experience. When user experience, the technology and the business model come together, it is possible to achieve what, in the innovation world, is called an 'iPhone moment' – the creation of a new product that becomes truly transformational. As I write this, within the sphere of AR and VR headsets, the technology is almost there, and so is the user experience, while the business models are pretty much there already. All we are waiting for is that little spark, where all three converge at the right time, to ignite mass adoption and propel the Metaverse forward.

Stepping into this brave new world

As a business, the best advice I can give you is to start conducting pilots with this technology to see how it can fit into your business and how you can use it to develop products that could be of interest to Gen Z and Gen Y, or even to refresh your customer base. Whether you want to improve the training you offer, or manufacturing and productivity, increase your brand and product awareness, or create completely new and completely digital products and services, you have to start using this technology. Step into the map and start making little forays to uncover the landscape around you. You never know, you might uncover some hidden treasure! And at least if you do run into an obstacle, you now know it's there and can start preparing to overcome it.

While this will involve a capital expenditure, you will build intellectual capital around the products you can bring over whenever you refresh the devices you use. You can create content for one type of headset now, and then, through smart, strategic choices around the platforms and infrastructure you use, you can have that content for a lifetime. As you use this technology to deliver training, whether it is just-in-time training for people in the field or induction training via VR, you'll streamline the way you work and, in turn, increase productivity. This will allow you to reduce costs elsewhere in your business. You'll create better products because your employees are better trained, and your employees will feel more satisfied in their roles because they know what they are doing and what is expected of them at all times. Workplace safety will also improve, with fewer accidents and injuries, as a result of training and in-built safety features.

From a consumer perspective, you can increase your brand awareness and reach an incredible number of potential customers from all over the world. Look at *Roblox* as an example – as of May 2022, there were more than 50 million active users per day on the platform. If you join *Roblox*, you can potentially reach out to those users and ensure your brand is seen by more young people. There are very few barriers to entering the digital goods business as compared to physical products, because the cost of producing 3D digital products and distributing them in the Metaverse is low, since these are digital in nature.

The key to being successful in the Metaverse is entering it with a strategy. You have to think very carefully about what you want to do. Start with a business goal, and set this as your North Star (where you would like to go, what you would like to achieve). You might also start by trying to solve a pain point. That pain point will be your starting point, where you can begin to design products and services based on technology enablers. Usually, if you have both a pain point (starting point) and a business goal (where you would like to go), then you have a clear pointed vector for your innovative thinking during the use case design phase. The use cases that emerge from this carefully aimed process are usually the most transformative.

Once you have your direction, you need to create a strategy for a pilot; a strategy for a minimum viable product (MVP – a product and service with the minimum specifications possible to achieve your objective); and a list of activities you could do to expand and build upon on what you create, also determined by the success of that first step (MVP). Your first step into the Metaverse as a business has to be measured based on key performance indicators (KPIs), which are defined during the design stage of your strategic plan.

Any ideas you come up with during the strategic assessment phase have to be prioritised by impact and feasibility. This is a solid strategic framework to design Metaverse use cases that could work for your business. In Chapter 8, I'll share more details on how to get started in the Metaverse as a business.

The most important part is to simply start. Start building a small MVP that can immediately create impact. Start creating your intellectual capital and developing your knowledge about the Metaverse. Start learning how to strategise in virtual worlds, virtual environments and the virtual economy. Start thinking about how to diversify your product lines and the possibilities the Metaverse holds for your brand. One piece of advice is not to limit your activities in the Metaverse to what you do in the real world. Look for ways in which you can branch out and build your offering in these virtual worlds in a way that complements what you offer in the physical world.

The Metaverse offers an incredible opportunity for companies to diversify their offerings far beyond their current products. Just as an individual can be whoever they want to be (non-binary unicorn doctor, anyone?!), so too can companies be whoever they want to be in the Metaverse. It is a place where businesses can explore multiple potential identities without risking too much in the real world. The Metaverse is ours to shape. It's time to step into this brave new world and potentially carve out a new identity for your brand and business.

Endnote

1. Carrel-Billiard M, Guenther D, Rosa N, and Taylor K (2021) 'Immersive Learning with XR', *Accenture*, available at: https://www.accenture.com/_acnmedia/PDF-164/Accenture-Immersive-learning.pdf

CHAPTER 7
SOCIOLOGY
OF THE METAVERSE

With contributions from Dr. Fadi Chehimi

Fadi has been in the immersive experiences space since 2004, when he developed novel concepts for mobile mixed reality, and he is now heading the global Consumer Metaverse offering lead in the Accenture Metaverse Continuum Business Group. Fadi has been instrumental to building Accenture's thought leadership and approach to the Metaverse, working with clients in a host of industries such as fashion, telco, government, banks, construction and media to help navigate their paths to success in this new spatial medium. Fadi is working with them on creating meaningful narratives for their brands, designing strategic value propositions for their business and defining the most optimal technical solutions to deliver. Fadi is a thought leader in the immersive experience design space. He has introduced a new design mindset for XR, 3D web and the Metaverse that moves away from focusing on 'users' to looking at them more as 'participants' in spatial experiences.

What do we mean by the 'sociology of the Metaverse'? In simple terms, the Metaverse is based on experience and social interactions, and what governs those social interactions is tribalism. These 'tribes' can be defined by specific activities, specific topics or even specific brands. In much the same way as we use different networks on the Internet to discuss everything from politics and music to events and business, the Metaverse will provide another space in which we can interact socially, yet in a more personal and immersive way.

The importance of tribalism in the Metaverse

As humans, we spend our lives aggregating around places where we can connect with other like-minded people. We have an innate desire to come together around certain topics, missions and values. In the past, we always needed a physical place to host these gatherings. We have seen in recent years that these places can be transferred online, whether through social media, gaming, video conferencing or now the Metaverse. What has not changed is this human need to be sociable. We are social animals by nature, and that is why it is so important to consider community in the Metaverse, particularly if you are a business that is looking to carve out a presence there.

What made social media so successful was the social element of what it offered (the clue is in the name!). We cannot expect to move into the next iteration of the Internet without inheriting and building on top of that social element. People will want to come together, form connections and build communities on the Metaverse just as they do in the real world.

Tribalism comes into play when we start defining the familiar characteristics within those communities, such as age, gender, background, values and so on, and the new ones for the Metaverse, like community affiliation, NFTs in a digital wallet, virtual spaces visited, style of avatars befriended, etc. We can find a common topic at the heart of any community, and it is this that brings people together. As humans, we want to belong to something and, once we are part of a community, it is very difficult to leave. Brands and businesses looking to enter and participate in the Metaverse need to understand this concept if they are to be successful. When brands understand this, play within these rules, speak the same language and share the same values as a community, they will become part of that community. This enables far more meaningful interactions than simply bombarding people with sales and advertising messages.

Adidas has demonstrated how to do this well through its partnership with the Bored Apes Yacht Club (remember them from Chapter 2?). Adidas understood the value of this community and tried to play by their rules and social constructs. The brand bought some of the Bored Apes Yacht Club NFTs and have since dressed their Apes in Adidas clothing and created unique personas around them. Rather than forcing the business directive on the community, Adidas rode the wave, and that has allowed them to join the community and make valuable connections.

Two decades ago, it was common to say 'content is king' to focus businesses on what attracts eyeballs and drives traffic. A decade later, it was all about how business products and

services looked and felt to simplify people's interactions and lead to faster transactions. That is when 'experience is king' succeeded to the throne. Now, with the advent of new expectations and social desires in the Metaverse, tribalism comes to build on top of content and experience to deliver intimate human needs. As Fadi puts it, in the Metaverse, 'community is king'.

The power of social tribes in the Metaverse

As with other parts of society, we are already seeing communities of people within the Metaverse who aggregate around sports, others who aggregate around celebrities and others who connect over events and music. There are huge opportunities for brands and businesses who take notice of this movement and create a strong presence there, especially in these early stages of the Metaverse's development.

Case study: From open mic to investment

For example, a 16-year-old girl has managed to secure $20 million in investment to build a business based on giving people a community space where they can perform their music and share their art on *Roblox*. She started out by creating her own virtual identity, an avatar called 'Kia', and by playing her electronic music through her avatar on *Roblox*. She started amassing fans, and that led her to hosting events where both she

and her fans could perform music in this virtual world. Simply by consistently hosting these events and visiting these virtual spaces, she has built a following and a community of 300,000 people, which enabled her to convert this hobby and the community into a multi-million-dollar business[1].

Another great example comes from Gary Whitta, a well-known video games journalist and former editor of both the UK and US editions of the *PC Gamer* magazine. During the COVID-19 pandemic, he created his own talk show, *Animal Talking*. The show was entirely live streamed from a talk show set he built in the game world of *Animal Crossing: New Horizons*, and it was streamed live on Twitch, with highlight videos also made available on YouTube.

The show involved interviews with guests who not only talked about their own lives, but also about *Animal Crossing*, as well as live music performances and stand-up comedy. The show ran for two seasons with a total of 26 episodes and was incredibly successful. By its fourth episode, it had 12,000 viewers, and episode 10 attracted a total of 339,000 individual viewers. Episode 11 set a record for the show, reaching 18,000 concurrent viewers[2]. What Gary Whitta did exceptionally well was bring together the tribe of people who played *Animal Crossing* by creating content via the platform that they wanted to see.

By hosting the talk show from a set built in *Animal Crossing*, he made himself present in that space and demonstrated he

was part of the *Animal Crossing* community, and that he understood their values, which contributed significantly to his success with the talk show.

Another way in which the real world and our virtual worlds are blending is through the likes of virtual music concerts. *Fortnite* has been experimenting with this concept and, in 2020, took it to a whole new level with its Travis Scott tour. This meant the artist played a series of five gigs in the platform, all live and with spectacular effects to accompany each song[3]. His *Fortnite* performances were attended by 12.3 million people[4] – where on Earth would you ever have a real-life gig with that many people attending?

This just goes to show what can be achieved with an audience you already have in place on an existing platform (or within a game). *Fortnite* players aren't there to attend concerts; but, once an event like this happens, news quickly spreads and people flock to the platform to check it out – just like you would head to your local park if you found out there was a free or impromptu gig happening (especially if one of the world's biggest music stars was performing).

The dark side of tribalism in the Metaverse

We have to be aware, of course, that there is a very real danger that these platforms can be used to influence and indoctrinate people into communities that are fuelled by hatred and are not acceptable in modern society. This is why it is so important for platforms to be vigilant of any form of hate speech and for governments to formulate regulations to prevent and tackle such instances.

There have already been examples of this happening in some of the early Metaverse platforms, like the rise of neo-Nazi and fascist groups on *Roblox*. In one scenario, a group of teenagers created a Roman-inspired game that had ever-increasing Nazi and fascist overtones and attracted thousands of players before it was removed from the platform in 2015[5]. This is particularly concerning because the people who typically engage with and play on *Roblox* are young and very easily influenced.

This is not a problem that will be confined to a single platform, and it is important that we be vigilant about this threat in the future as the Metaverse expands. It is going to be a very vast place, and policing what happens in the Metaverse will not be easy, so we need to consider how we can moderate and patrol these virtual environments now to prevent activities like what happened in *Roblox* from repeating elsewhere. Companies that are creating platforms that will form part of the Metaverse need to have automated systems in place to carry out this moderation.

Although governments need to create comprehensive governance for the Metaverse, organisations operating in this space cannot wait until this is put in place. They need to be responsible and take action now to send out a strong message to the members of those communities that are spreading hate speech and fake news – that there is no place for them in the Metaverse.

This argument, of course, overlaps with questions about freedom of speech, but it is important to be mindful that certain topics should not fall under the gamut of free speech, especially in virtual environments where there can be such large echo chambers and audiences that are so malleable. There need to be some limits imposed around the discussion of topics such as child pornography, domestic violence, hate speech and even modern digital slavery, for instance, to limit extremist ideology being spread on these platforms.

We have to make sure there are checks and balances in place to prevent fringe groups from spreading hate speech and political messages that are not ones of inclusiveness and peace. If we don't, these platforms risk becoming places where those around the world who have extreme and potentially dangerous ideals can indoctrinate and recruit others to their cause.

We have an opportunity to bring the world together in a more understanding and cosmopolitan way, because you will find people from all over the world in the Metaverse, and people will be able to meet each other independently regardless of where they live. They can talk together, play together and work together towards their shared interests, which is wonderful and beautiful. This needs to be protected against the actions of those in society who seek to divide and cause conflict.

The safety of your senses

The Metaverse will be more than just a platform you visit via a mobile phone or a browser on a PC. It has the potential to be a fully immersive environment, and that means all of your senses will be involved. If you are visiting somewhere in the Metaverse and you are harassed, it has the potential to become very emotional, very quickly, because all of your awareness is in that space, from your spatial awareness to your sight, hearing and potentially even touch. The Metaverse isn't like social media; you won't be able to just scroll past threatening posts. You are there, inside it!

This means that, if you are harassed in this immersive virtual world, it can be a very scary experience.

'My kids, who are nine and six, have started to explore Roblox, and they each have certain games they like to play. One day, my son saw an ad for a game that features a rather horrifying face. The concept of this particular game is that you have to run away from the horrifying person so that they don't hug you. Bear in mind that my son only saw the advert – he didn't play this game – but he was terrified. All of a sudden, not only did he not want to go into Roblox any more, he didn't even want to play on his iPad. You cannot underestimate the impact that these experiences can have on people of all ages'. Fadi

The example Fadi shared about his son did not involve anyone behaving inappropriately; it was simply that the child was scared of something he saw in relation to content on the platform. However, there have been incidents where people have been physically intimidated and scared in a virtual environment. In one such incident, a blogger reported on *Medium* that she was sexually harassed by three to four other players with male avatars on Meta's *Venues* platform. They virtually gang-raped her avatar – an incident she was forced to watch – and then sent her highly inappropriate comments and photos they had taken of the virtual assault. She was shocked, and the incident left her feeling unsafe and uneasy. It was an emotional experience for her, even though it didn't happen in the real world[6].

Meta have taken note of this incident and have implemented a feature on their platform that prevents other avatars from invading your personal space. And, if they are allowed in, the hands of the avatar will vanish in a measure to tackle harassment. Much like the option you have to engage in player-versus-player (PvP) combat in *World of Warcraft*, this gives you the opportunity to choose who you engage with.

Being able to have anonymity behind your avatar may currently embolden some users to take actions they would never take in the real world or in a virtual environment when they could be identified. However, it is likely that, at some point, there will be a way to track everyone's digital identities, which will hopefully act as more of a deterrent towards those considering behaving in inappropriate or threatening ways towards others in the Metaverse.

Roblox is now one of the better platforms in terms of having mechanisms in place to police behaviour and allow people to report inappropriate behaviour. Players can also vote to keep other people out of specific spaces. However, *Roblox* started in 2008 and has had a lot of time to develop these parameters and find the most suitable ways of moderating its virtual environment.

Virtual reality and responsible immersive experiences

Virtual reality (VR) is one of the technologies that will be used extensively in the Metaverse, and it is also one of the most immersive technologies we have available at the time of writing. The more immersive the experience, the closer it is to reality, which can be a blessing and a curse.

Studies have shown that VR can be used as part of a very effective therapy to treat people with post-traumatic stress disorder (PTSD), which work by allowing people to relive an experience in a safe environment, sometimes with the support of medication, and, in doing so, help them realise that it is not as harmful as they might think. The same technique can also be used to help people combat their phobias, whether they have a phobia of flying or spiders. By being exposed to these situations in a safe environment, that person creates a neural pathway that helps them deviate from the anxiety and stress they feel. This neural pathway is strengthened in the VR environment, and then can be used in the real world as well.

However, while VR can clearly be used to help people, it can also be used for darker purposes. We need to be particularly mindful of this darker side, and be aware of the influence that VR can have, given the rate of development within VR technology. Basic VR headsets, at the time of this writing, give you spatial awareness, sound and vision in a virtual world. From 2022, headsets will be available that have eye-tracking technology embedded in them (the first ones are due to be released by Meta and Sony). These headsets will not only track where you are looking, but also the dilation of your pupils and a great deal of additional biometric data. This will therefore allow whoever has access to that data to moderate the experience to instil certain emotional messages within the content you see. This could, and very likely will, be used in the future for advertising purposes, or potentially misused to spread political messages based on the user's emotional response to audio-visual stimuli.

We know that fear has long been one of the most powerful emotions used in political campaigns to encourage voters to choose one person or political party over another. God only knows what could be possible if some of the political leaders who focus on fear-based messaging for their campaigns were able to share those messages via VR in the Metaverse. It is important to be aware of the potentially disruptive impact this technology could have on people's lives.

These immersive experiences can feel so real that they can be incredibly intense, and we need to be mindful of this.

> 'One of the most powerful experiences I've had in VR is a simulation of an atomic bomb explosion. You see a three-dimensional nuclear explosion recreated before your eyes. You see the impact of the shockwave around you, with trees shaking, and everything around you catching fire. I had goosebumps when I experienced this in VR; it was one of the most shocking experiences of my life'. Nick

The atomic bomb explosion experience could be great for education and awareness in the right context, but we always need to be mindful that VR emotional impact is by far more powerful than any previous media we have had access to date. This won't be an appropriate medium to use to raise awareness of the fallout from a nuclear attack among everyone. Some people might see it as an enlightening experience, whereas others may find it too intense and scary. We need to consider how those providing VR content on platforms within the Metaverse can communicate the intensity of the experience to users to help them make the decision about whether they want to engage in that experience in the first place.

We have a rating system for films and games to help us decide what content is appropriate for ourselves or our children. It could be that parts of the Metaverse carry similar ratings to help all of us make choices about where to go based on what we feel able to tolerate in a 3D immersive setting.

First and foremost, it is the platform's responsibility to look after its users, and these are the kinds of things platform developers need to consider. However, there will no doubt also need to be some form of government regulation, although this can lead to many grey areas.

Imagine that a 14-year-old girl goes into a virtual pub in the Metaverse and starts consuming alcohol in this virtual environment, essentially exposing her to the concept of drinking at a young age. Who is responsible for that breach? Is it the girl who went to the pub, even though she knew she was underage and not allowed in that environment? Is it the owner of the pub? Is it the platform that allowed her to get to the pub in the first place? Or is it the government that made the rules?

The challenge with a global network of online environments like the Metaverse then becomes, which government has jurisdiction here? Is it the government of the country the girl lives in, the government of the country where the pub owner is based, or the government of the country where the platform is based? It will become very messy very quickly. There is no escaping the fact that governments will have to play a role somewhere down the line, but, due to the global nature of the Metaverse, we have to consider how governments can reach a global consensus on issues such as this.

However, the challenge at present is that the Metaverse and the technology behind it are all so new that many governments haven't even thought about these particular issues yet. This is why the companies building and operating platforms in the Metaverse have an ethical responsibility to design everything in a responsible way and consider these potential issues. This is true of platforms, technical providers, vendors, system integrators and consulting agencies. We all need to design in a way that is responsible and inclusive, while always considering the impact that everything we build could have on society.

We can no longer just drop technology and ideas on the world because we think they're cool, which is what happened in the early 2000s. We have all seen how certain social media platforms have polarised the world and even been weaponised by governments and others due to a lack of any meaningful oversight or regulation. We cannot allow this to happen again.

Data ethics in the Metaverse

Everybody operating in this industry has an ethical responsibility to regulate the Metaverse properly and avoid any scandals, such as what happened with Cambridge Analytica. This is not only about being responsible for the platforms themselves, but also about being responsible with people's data and making sure that it is used in the right way.

The Metaverse is not only incredibly immersive, but also has the greatest potential of any existing medium to profile users in a very deep and intimate way. Mass data collection targeting users of the Metaverse is a significant threat and one we need to consider and mitigate now.

The array of data you can collect using augmented reality (AR), virtual reality (VR) and other immersive technologies is vast. It is not only about what people say, but also the way in which they speak, and biometric data including eye gaze, facial expressions, pupil dilation, movement in space, body posture and more. It is possible to collect millions of intimate data points from people in the Metaverse. When this is combined with technology, it has the potential to produce believable virtual avatars who can talk to you and convince you to do anything they want because they know you so deeply.

At the time of writing this book in 2022, we see an advert and click on it because companies are profiling us based on our online browsing behaviour to serve us ads that they know we will be interested in. These companies know if we're searching for a car, looking for a new house or thinking of buying a new synthesiser. They know this because they track our interactions, they listen to what we say and they build a profile of us. Now just imagine the power these companies could have over you if they knew exactly what triggers a particular emotion for you. This data could be weaponised in so many ways, from encouraging you to buy a particular product to making you believe a particular political message. We need to put safeguards in place to ensure this does not happen in the Metaverse.

There are three main safeguards that are needed to steer us away from this terrifying vision of the future of the Metaverse:

1. Place limits on the kind of profiling that companies are allowed to carry out in the Metaverse.

2. Prevent companies from carrying out any kind of emotional profiling of users.

3. Prevent biometric data from being stored for any advertising or non-health-related reasons.

Louis Rosenberg, technology pioneer, chief executive officer and chief scientist of Unanimous AI, has long been calling for these three pillars to be enacted across the Metaverse[7]. He's considered one of the inventors of AR and is a luminary in the AI industry, giving him an incredible level of understanding about how this technology can not only be used for good, but also how it could be manipulated or weaponised to the detriment of all of us.

Community influence in the Metaverse

We also have to consider how the communities or tribes we join and build in the Metaverse can influence both ourselves and others. The desire to fit in with those in your tribe doesn't disappear in an online space, so if everyone else's avatars in your Metaverse community are dressed up in Gucci, you are likely to want to do the same. This kind of social conformism is nothing new. For brands, however, it presents an opportunity that should not be overlooked. You can build communities where you are able to showcase the value of your product not through advertising, but by social acceptance.

However, we also have to be mindful of class mechanics in the real world being replicated in the Metaverse. When you look at the Metaverse as a copy of society, where all the same rules around wealth creation apply, it is easy to see how this could simply result in the rich becoming richer. There are many examples where NFTs have sold for millions, which is obviously out of reach for most people. Bored Ape Yacht Club NFTs, for example, hit a record high average price of $308,497.23 by April 2022[8]. However, this perspective ignores the fact that the Metaverse is *not* a carbon copy of the real world, and that, as it evolves, we are going to see more and more use cases and opportunities developing.

What we have in the real world is a centralised economy, whereas what the Metaverse offers is a decentralised economy. In a centralised economy, if you have an idea for a product, you need the money to develop and manufacture it. In the Metaverse, this isn't the case. You can do everything digitally, which is easier, faster and allows for mass production without needing a substantial capital outlay.

This can be used to help redistribute wealth around the world, because some of the money that has been tied up in the physical economies of the world's wealthiest countries can be spent in places where people have less. Your physical location does not matter when it comes to creating products or earning money in the Metaverse. We are already seeing huge tribes of people in the Philippines, for example, who make money by playing crypto games. This kind of wealth redistribution will continue as the Metaverse expands. One of the great things within a global, decentralised economy is that it will remove the need for exchange rates and bank transaction fees. We are looking at an entirely new model for earning money and doing business.

What is the meaning of reality?

Think about our reality right now. In this world, if you leave your mobile phone at home, you probably feel anxious and want to go home to get it. If your phone breaks or gets stolen, you probably feel disconnected from the world, and your reality becomes less rich. It is as though your superpowers have been taken from you – all of a sudden, you can no longer hail a taxi, message your friends, call your lover or join a work meeting on the fly. In the future, this distinction between having and not having technology is going to become even more marked.

Once smart glasses become ubiquitous, if you lose them or you don't have them, not only will you be disconnected in a similar way as losing your phone, but you will also miss out on the richness of the reality that surrounds other people, because those who wear smart glasses will see things that people without smart glasses simply can't. Those with smart glasses will live in a richer reality, whether we are talking about how they receive information, the entertainment they access or even communication. There are many philosophical arguments around this, which are covered extremely well in David J. Chalmers' book *Reality+*.

We have to be mindful that we could create a huge digital divide between those who can afford to have smart glasses and those who can't. We also have to consider that, in the future, virtual worlds and virtual assets will be as real to us as the physical world and our physical assets, and therefore people who are deprived of access to that virtual world will live a life that is not as compelling or rich as the lives people live when they can pass seamlessly between the virtual and physical worlds.

The way we perceive the world and reality as we know it will shift dramatically in the next 20 years. Right now, we don't know the answers to all these questions or how our perceptions of the world and reality will change. If we put ourselves in Plato's Allegory of the Cave, we are still very much in the dark. We are deep in the cave, and we are only seeing shadows of what might be beyond. Right now, those shadows are of a world outside the cave where we are using smart glasses to access the Metaverse and to move between that and the real world; but, when we actually step out of the cave, this might look different to how we imagine it now.

Returning to the idea of redistributing wealth and everything being possible in the Metaverse, we also have to consider the potentially negative impact this can have on communities. In the Metaverse, it may be possible to have assets that are status symbols, like a Gucci bag or a Lamborghini, that we would never be able to own in the real world. In the Metaverse, those status symbols might cost just a few dollars. However, if we get used to the idea that everything is possible (and this is likely to be a particular challenge for younger generations), and valuable assets are devalued in our minds by what we see and experience in the Metaverse, we are creating a conflict between what people are used to having and doing in the Metaverse and what they have and do in the real world.

In a virtual environment, we may behave differently to how we would in the real world. However, as we spend more and more time in those virtual environments, which behaviours become more prevalent? If you behave in a way that is socially unacceptable in the real world, but which is accepted in the Metaverse, will you bring that social misconduct into the physical world with you? Will there be a difference in how people deal with and interact with avatars as opposed to real people? Will the social norms we are used to in the real world be different in this virtual world, and what will that mean for our behaviour in both?

We also have to be mindful that, in a virtual world, we could have multiple avatars with multiple identities or personalities. This will allow us to take on different roles and behave differently depending on which of our identities we assume. However, this could lead to a situation where we become unsure which of these identities we should assume and use in the real world. If you look at social media in the world today, you can already argue that we all have different personas on different platforms. Our LinkedIn persona will be different from our Facebook persona, which will be different from the persona we display in Instagram and different from our TikTok persona.

The social environment we are in dictates our behaviour. We even do this in the real world – in that the way we behave with our friends is different from the way we behave with our colleagues. However, in an immersive environment where we have the opportunity to have different avatars and identities for different situations, this could become more pronounced and also influence the way our personas develop, particularly if we are using this technology from a young age.

Crime in the Metaverse

Of course, this also opens up many other areas of debate, such as how crime in the Metaverse will be treated. Will crimes in the Metaverse be treated in the same way as crimes in the real world? This is a vast topic, and one that is far beyond the scope of this book and my specific expertise. However, it is still important to consider, particularly in relation to the ethics of the Metaverse.

At present, under international legal frameworks, if a crime is committed on the Internet, then the person in question is judged by the authorities in the country in which they are residing and therefore from where they access the Internet. So, if I am connected to the Internet in Spain and I commit an online crime, I will be charged and prosecuted under Spanish law. In the Metaverse, these rules would also apply as things stand. There are also certain crimes that a specific platform would be obliged to report under international law, such as child exploitation. In this case, if a platform uncovered evidence of child exploitation, or a user reported it to them, they would be obligated to report that to the local authorities, again based on where the user committing the crime resides.

It is very likely that this will be a standard model that Metaverse platforms adopt going forward. Each platform will need to set its own regulatory standards, rules, and terms and conditions (T&Cs) that set out how and when they will report incidents or crimes to local authorities. In fact, the Metaverse platforms that are operating now have an opportunity to create a gold standard that can be held up as the standard to follow by any that follow them.

We are already seeing evidence that governments are seeking stricter regulation of social media platforms, and it naturally follows that this will transfer to the Metaverse as well. In April 2022, the EU agreed on its landmark Digital Services Act, which will hold online platforms accountable for any illegal or harmful content published on their sites, even if it is user generated. Once it is adopted, by 1 January 2024, it will mean platforms can face hefty fines for failing to tackle hate speech, the spread of disinformation and fake news, and the sale of fake products[9]. While developed with a view to reining in the world's social media behemoths, it will be just as applicable, and arguably even more important, for driving regulation across Metaverse platforms.

There is always a bright side

Like every medium, there is no way to eliminate all of the bad and only have the good. There will always be someone who will exploit it for malicious purposes. While we can't stop that from happening, what we can do is reduce the likelihood of this having far-reaching consequences if and when it does by making those joining and interacting in the Metaverse aware of the dangers for potential misuse, while designing these environments in a safe and secure way.

This book also hopes to be an ethical and business guide to the Metaverse, because this virtual environment is not just about having fun and playing games while making money. It has the potential to be highly disruptive technology, and this means it could be manipulated or even weaponised. One of the aims of this book is to ensure we have conversations about how we can avoid the Metaverse being used in a malicious way and, instead, ensure we are able to enjoy all the fantastic opportunities this new virtual world offers.

The Metaverse will give us the ability to freely express ourselves. All of us will have the opportunity to belong to communities we care about and to be influential in those communities. It will open up new opportunities to co-create and design the next generation of products or services, all of which will enhance and add value to our digital selves. This can empower us, both in the virtual and real worlds.

As a short example of how this could positively impact our lives, a TV talent show in the United States called *Alter Ego* gives those artists who do not conform to the classic idea of beauty, have disabilities, are not young anymore or who simply have a fear of standing in front of a crowd the possibility to perform using a 3D digital avatar and a full motion capture suit. This means it is not them who gets up on stage, but an avatar of themselves that they have created, one that represents them and performs on stage on their behalf[10]. This is not replacing oneself, rather augmenting it.

The Metaverse presents a fantastic opportunity for inclusivity, but in order to be truly inclusive, it is essential that these digital worlds be designed with disabled people in mind. We have to consider what alternative technology could be used to accommodate those with physical disabilities. For example, how can you use virtual hand tracking with someone who is missing that limb? How will the Metaverse translate for those with sight or hearing difficulties? These are all questions we need to ask now, so that we can ensure we create a Metaverse that is truly accessible.

The Metaverse has the potential to be an incredible tool, but we have to be mindful of the divisions it could create within society and explore how we can reduce or eliminate those, so that we can create a Metaverse that is inclusive, ethical, responsible and wonderful for all to enjoy. We have the power to change the world – let's use that superpower for good.

Endnotes

1. JR (2021) 'Splash raises $20M to expand Roblox music-making tools', *JackOfAllTechs.com*, 2 November, available at: https://jackofalltechs.com/2021/11/02/splash-raises-20m-to-expand-roblox-music-making-tools/

2. 'Animal Talking with Gary Whitta', *Wikipedia,* 27 March 2022, available at: https://en.wikipedia.org/wiki/Animal_Talking_with_Gary_Whitta

3. Webster A (2020) 'Travis Scott's first Fortnite concert was surreal and spectacular', *The Verge,* 23 April, available at: https://www.theverge.com/2020/4/23/21233637/travis-scott-fortnite-concert-astronomical-live-report

4. Amore S (2020) 'Travis Scott's "Fortnite" Concert Attracts Record 12 Million Viewers', *The Wrap,* 24 April, available at: https://www.thewrap.com/travis-scotts-fortnite-concert-attracts-record-12-million-viewers/

5. D'Anastasio C (2021) 'How *Roblox* Became a Playground for Virtual Fascists', *Wired,* 10 June, available at: https://www.wired.com/story/roblox-online-games-irl-fascism-roman-empire/

6. Shen M (2022) 'Sexual harassment in the metaverse? Woman alleges rape in virtual world', *USA Today,* 31 January, available at: https://eu.usatoday.com/story/tech/2022/01/31/woman-allegedly-groped-metaverse/9278578002/

7. Rosenberg L (2022) *Should You Fear The Metaverse? VR Pioneer Explains | The Classroom,* More Perfect Union, YouTube, available at: https://www.youtube.com/watch?v=9HvqIZvhZ3M

8. Minter R (2022) 'Bored Ape Yacht Club Surpasses $2 Billion in All-Time Sales', *be in crypto,* 5 May, available at: https://beincrypto.com/bored-ape-yacht-club-2-billion-sales/

9. European Commission (2022) 'Digital Services Act: Commission welcomes political agreement on rules ensuring a safe and accountable online environment', press release, 23 April, available at: https://ec.europa.eu/commission/presscorner/detail/en/ip_22_2545

10. Hissong S (2021) 'On the Set of Fox's Trippy New Reality TV Show, Which Turns Singers Into. . . Aliens?', *Rolling Stone,* 21 September, available at: https://www.rollingstone.com/pro/features/alter-ego-singing-competition-augmented-reality-tv-show-virtual-avatars-1228045/

CHAPTER 8
SUGGESTIONS FOR A STRATEGIC APPROACH TO THE METAVERSE FOR BUSINESSES

With contributions from Maria Mazzone

Maria Mazzone is a Managing Director at Accenture, focused on innovation and industry. She has been leading the Accenture European Innovation Center for years, and has an Innovation Director role on many consumer goods and retail clients. She holds a Master in Anthropology, which, coupled with her work in innovative technologies and experience design, positions her uniquely to work with the Accenture Metaverse Continuum Business Group to chart a future path into the Metaverse. She lives in Milan with her husband, two kids Ludovico and Lavinia, and Luna the dog.

The Metaverse presents an opportunity to conduct business differently. We have so far discussed what could be described as a socialist utopia, whereby the economy within the Metaverse is based on what people are creating and building within this digital world. This is certainly one area where businesses can become involved and take advantage of the talent that's available, by providing these creators with access to their intellectual property (IP) and allowing them to use this to create products and services within the Metaverse. This is a process that can either work by moving top-down or bottom-up, making it more versatile than the economy in the physical world as it stands today.

One question that we find companies consistently seeking an answer to is, 'Who will rule the Metaverse?'. The challenge is that many businesses are viewing the Metaverse through the same lens as they view many areas of the tech industry, where there are one or two big players that have dominated a space, whether that is Meta and Twitter in social media, or Amazon and Apple in retail. However, there is no guarantee that the Metaverse is on the same trajectory as other sectors in the tech industry. In fact, there are indications that it will develop very differently to the tech economy as we currently know it.

Decentralised Metaverse platforms (like *Sandbox* and *Decentraland*) might be there to be purchased by a big tech giant. Many people who have been early adopters of these platforms and bought land within them would likely push back against any attempt at being bought out by one of the big tech giants. In addition, because these decentralised platforms do not have just one owner, it will be much more difficult for a big tech business to swoop in and buy the platform out. The people who make up that community will all have to, as individuals, agree to selling their land to this buyer, and finding this level of consensus across this many people will be a significant challenge for any company that wants to go down that route.

In all honesty, at this stage in the development of these decentralised platforms within the Metaverse, no one knows how a buyout could be approached successfully because, at the time of this writing, this hasn't happened yet. A good, if slightly simplified, way to think of these decentralised Metaverse platforms is like an apartment building. If you had 20 apartments within that building, each owned by a different person, then anyone who wanted to buy the whole building would need to get the agreement of each individual owner. Similarly, any decisions about how to manage communal spaces or the upkeep of that building would need to be agreed on by all the owners. In this situation, there is not just one person who makes decisions.

It is similar in the decentralised Metaverse platforms, where there are still discussions about how they will be regulated and how the owners of land within those platforms will manage them. Another crucial point that still needs to be defined, which is particularly important for businesses looking to invest in land in these platforms, is that of identity and there being transparency over who owns what.

As a business, in particular, it is important to know who or what entity is behind the majority landowners, shareholders or voters in a decentralised Metaverse platform. There are several reasons for this. The first is competition – for example, if a large multinational company wanted to buy some land but it belonged to its main competitor in the market, that could have serious implications. The second is to ensure that you are not funding crime, terrorism or other unethical activities. The last thing you want as a business (or an individual) is to buy land in a decentralised Metaverse platform only to learn the previous owner was a drug lord or terrorist organisation and that you have inadvertently funded their activities.

As a result, regulation will need to be introduced to protect everyone's reputations, allow businesses and individuals to behave ethically, and to prevent activities such as money laundering. This is, once again, where blockchain technology will prove crucial.

There are also some question marks over the value of land within Metaverse platforms that are basing their models on scarcity. These platforms are currently saying that there is X amount of land, and that there won't be an opportunity to build more land once it has all been sold. This means if, as a business or brand, you don't buy land in these platforms now, you will have to buy it from existing owners, which will drive the value of that land up, just like in the real world.

However, the question no one has answered so far is what will happen when we achieve interoperability between these platforms and the various platforms that form the Metaverse are interconnected? The value of the Metaverse platforms where land is limited will essentially be diluted by the fact that we, as users, are able to travel from platform to platform so easily that no one can tell where one platform ends and another begins.

There are also queries around non-fungible tokens (NFTs) and how these will be managed in the future, because, although an NFT is a digital asset, it still has to be hosted somewhere, and that hosting service will need to be paid for. There is no clarity over whether the company that sold the NFT should cover that cost, or the person who bought it. If you pay $20 million for an NFT of a digital image, and the cloud service hosting that digital asset expires, you will be left with a broken link and no way of seeking recourse because there is no regulation in place to cover this at the time of writing.

Many individuals and businesses are part of a frenzy of buying in the Metaverse in this early stage, whether we are talking about land on Metaverse platforms or NFTs associated with digital assets, but these questions will need to be answered sooner rather than later. If you are a business that intends to offer NFTs, it is sensible to think about how you might tackle issues like paying for hosting and how you will manage that in the future, as well as how you will communicate the details about this aspect of NFTs with those buying them from you.

How will the Metaverse impact businesses?

The impact the Metaverse will have on businesses of all sizes will be huge, but it will affect small, medium-sized and large multinational companies differently.

For large multinational companies, the biggest opportunities come from the size of the audience they can reach. For example, Galleria Vittorio Emanuele II is a famous luxury shopping street in Milan. Before the COVID-19 pandemic, it was estimated to attract 22 million visitors a year. At the time of writing this book, *Roblox* attracts 50 million visitors *a day*. The number of people that brands can reach in the Metaverse is infinitely higher than any physical location. Nike is one brand that has moved into the Metaverse. It opened NIKELAND in *Roblox* in mid-December 2021, and by March 2022 had recorded almost seven million visits[1]. There are no physical stores that attract that kind of footfall in such a short space of time, especially following the pandemic.

There may be questions about whether the people visiting these Metaverse stores are the 'right' visitors with the kind of buying power that brands are looking for, but this misses the point. Brands need to work out how best to turn those visits into paying customers, whether that is by appealing directly to the demographic visiting them in the Metaverse, or by encouraging younger Metaverse users to bring their parents (who have the buying power) to these Metaverse-based stores.

However, the greatest opportunity for large businesses is to pivot their business model from what they are now to what they hope to be in the future. How a business can pivot will naturally depend on the industry they're in and the model they currently operate. If we look at a large consumer goods business as an example though, we can see just how significant the Metaverse could be in terms of how they do business.

Businesses in this industry have always operated on a product-centred model, where they make a drink or food product, advertise that product in the right way and work with retailers to sell and distribute it. In the Metaverse, businesses like this have the opportunity to pivot from offering a product to offering an experience. To do this effectively, the business needs to look at the core of what it's selling in order to find the most appropriate opportunity. For example, a company selling energy drinks could pivot to become an online wellness and fitness platform; a company selling alcoholic beverages could set up an online bar or even a dating service; or a beauty products retailer could become a beauty salon and a place where people go for advice and makeup tutorials.

Many consumer businesses have been trying to evolve from selling a product to selling an experience for many years without success. In many cases, they have tried to go down a digital route in the past, only to find that digital channels simply weren't strong enough – but that is all changing with the arrival of the Metaverse. The Metaverse is different to digital, and the best way to describe it is that, until now, operating digitally has been the equivalent of allowing someone to walk down a physical street and window shop without going into any stores. With the Metaverse, we are opening the doors to those stores, allowing consumers to step inside and giving them an all-encompassing experience.

We all know that window shopping, where you're standing on the outside looking in, is vastly different from going into a physical store where you can experience the vibe and touch the products – even the smell will be different. This is what the Metaverse offers and why it will be game-changing for so many businesses.

It isn't only consumer goods companies that have the opportunity to pivot in the Metaverse; even companies selling basic services like energy or telecom can carve out their own niche in the Metaverse. For example, an energy company could create a platform about green energy and become a place where children go to learn about sustainability. This is not simply about finding a new way to sell your products or services, but an opportunity to become a beacon for what you and your business stand for. These opportunities are everywhere – they are just waiting to be discovered on your greyed-out map of the Metaverse.

Creative collaboration is the future

Businesses also need to resist the temptation to simply think about how they can transfer what they do in the real world into the Metaverse. If you are a fashion brand, moving into the Metaverse is about more than just deciding that you're going to have fashion shows in the Metaverse instead of on a physical catwalk. That's great, but you also need to consider how you can bring consumers and others into your creative process. We're talking about taking a step towards a more open-source, collaborative environment.

Let's stick with the fashion example. Rather than launching a collection and just hoping people like it, you could invite your audience, buyers and ambassadors to contribute to your creative process. You might take suggestions on materials to use, colours to incorporate or even design features.

Through creative collaboration, there are also opportunities to build on what has already been created, thereby amplifying and multiplying the products that each company has. For example, a fashion company could launch a new line of virtual clothes for the Metaverse, with each item coming with its own NFT to create ownership of that particular virtual product. Now imagine that the owners of those NFTs are allowed to modify the item of virtual clothing, obviously within the parameters set out in the terms and conditions. This could mean your product multiplies, because different people will alter that clothing in different ways. Some might simply change the colour, others could tinker with the design – but, all of a sudden, your range has expanded. Some of those new products may even find their way onto the real-world market.

Eventually, one of those products will emerge to become more successful as compared to the others, because it is better. The revenue from this product will not only go to the creator, but also the company, ensuring that everyone who has had a hand in its creation receives a benefit.

The Metaverse, unlike the systems we have used to date, features a series of tools (that we discussed in Chapter 5) to allow companies to safeguard their IP and their brand. It is easy to attach certain conditions to an NFT that allow you, as a business, to revoke someone's use of that asset, should they infringe or breach those terms and conditions.

The world is your focus group

The Metaverse also allows businesses to completely change the way in which they develop and produce prototype products. We are talking about a fully digital process that does not require a manufacturing plant. In the old world, a team of people within a business would come up with a design for a new product, discuss it amongst themselves, make some changes and then take it to a focus group.

In the Metaverse, you can get input from your customers or potential audience much earlier in the process, and without incurring significant costs. You can create a digital prototype of any product in the Metaverse, which other people can view, move around and potentially even suggest modifications to. Unlike traditional focus groups and testing that relies on people being in the same physical location as the product they are testing, you can do all of this virtually in the Metaverse, and you can allow your designers or engineers to engage people participating in those focus groups in conversation about their thoughts. You could effectively crowd-source your product across the world while safeguarding your IP and with minimal risks.

> 'If I were given the opportunity to consult with any major car manufacturer, I would ask them to create a space in their cars for a handbag – because I've spent years putting my handbag on the floor, and it annoys me that I pay so much for a car and yet have nowhere suitable to keep my bag. In the Metaverse, I could work with a car manufacturer's engineers to find a place for this feature that works within the design of the car. We could test my suggestions on a digital model, moving it around until we found a solution. The speed of the impact you can have with this kind of research and development is incredible'. Maria

Similarly, the Metaverse opens up unrivalled opportunities to connect with people in different countries and communities. If you want to know whether a particular car feature works for women in Japan, you can easily find five Japanese women to participate in a focus group, for instance, all without the expense of travelling to Japan, transporting a prototype and running sessions in a physical location from the country.

Extended reach: a key opportunity for small and medium-sized companies

While what we have discussed so far in terms of the impact the Metaverse could have on businesses has focused on large, multinational corporations, that is not to say that small and medium-sized businesses cannot benefit in a similar way. However, the biggest opportunity the Metaverse offers smaller organisations is extended reach.

If you set up a business selling a new type of drink, for example, your market at present would consist of your neighbourhood. With low starting capital and low stock of the product, you become trapped in a catch-22 situation, whereby you need funds to expand your operations beyond your neighbourhood, but to get funds you need to have a much bigger market. In the physical world, it can take decades to build up a business by gradually making money, reinvesting that in the company and using it to grow.

By contrast, in the Metaverse, results and growth can be obtained much more easily with relatively small investments. If you were to set up your business in *Roblox*, you would have the platform's 50 million users as your audience. While you would still need some investment to get started, you would suddenly have access to a much wider market and, if your product is good, you can grow your business considerably faster than would be possible in the real world.

However, we would not recommend that small or new brands and businesses start out by purchasing prime real estate in the Metaverse. Having a store in the Metaverse as an unknown brand doesn't make the most sense. Instead, you can use the Metaverse to find your ideal audience within that space and build a strong community around your brand. This allows you to build up your following and find people who love you, so that you develop your presence and, in doing so, are more likely to attract investment.

Community and experience are key for thriving in the Metaverse

Not all businesses have a 'sexy' brand, even if they are large and successful. Not every business can be an international fashion label, a popular sports brand or Snoop Dogg. How can you use the Metaverse to your advantage if you're a company that operates in the energy industry or banking sector? One option, just like in the real world, is to sponsor and thereby attach the business name to events and/or personalities that appeal to a young and broad audience, like a virtual Justin Bieber or Deadmau5 concert. Another option is to attach your name to funny and engaging experiences and use these to create a community.

Imagine a multinational bank creating a virtual zoo, where people could learn about different animals, feed them and even help take care of them. They could even host competitions and allow people to earn virtual currency. All of this feeds into the experience you offer, as well as what the community is able to create around your specific brand. When that community reaches a critical mass, it will allow the business to thrive in the Metaverse and eventually to make some money.

The question that businesses must ask is, what experiences do people want to have? No one wants to spend time queuing in a bank, so simulating a virtual bank won't be appealing. Going to the virtual zoo, on the other hand, sounds like a great day out. Products in the Metaverse can be digital or experiential – but, to succeed, they need to be things that people want to experience or own.

Painting the world with augmented reality (AR)

Augmented reality, or AR, has the potential to turn any surface in the world into ad space, or to deliver marketing activation to consumers wherever they are, especially given that content can be shown to them based on their geolocation data. You can paint the world however you want. With companies like Google and Niantic having already created a pretty accurate digital map of the world, AR can really come into its own.

Using this digital map, it's possible to overlay AR content onto the real world relatively accurately, and this could be used for a range of purposes. For example, businesses could guide people through gamified experiences that are fuelled by to-kens, coins and offers that can be redeemable to customers in specific locations. As a business, you can place these collect-able items at strategic locations around the world. This could work particularly well if you were to position them in close proximity to points of sale that are connected to your com-pany and the experience you're offering.

You could also introduce triggers to encourage consumers to spend these points or tokens, or to use their offers, while they are in that location. For example, you might offer a 20 per cent discount on your products for collecting the tokens, but if they make their purchase in the next 15 minutes, they'll get 30 per cent off instead. The AR content can then signpost them to the nearest point of sale to further encourage them to make the purchase.

This is just one example of how AR technology could not only increase your organic footfall, but also create a sense of urgen-cy for consumers to visit a specific location and spend money there. There are many applications for AR, from restaurants being able to showcase their customer reviews and ratings to people while they're walking down the street, to playing games that blend the digital and real worlds.

Pokémon Go is an excellent example of how games in particu-lar have evolved as AR technology has improved. When *Poké-mon Go* launched, you would see your Pokémon appear on your mobile phone screen, a bit like a sticker that had been stuck on top of the real world. Now, however, all the 3D data we have about the world means that digital data can interact with the environment. Your Pokémon is no longer just floating on your phone screen, but can hide behind real-world objects like trees. The only way to see that character is to move around the tree. Just imagine the hilarity of chasing Pikachu or Squirtle round a park, seeing these characters peeking out from behind a hedge or under a bench! Using this kind of mixed reality approach (digital objects interacting naturally with the topology of the real world surrounding the user instead of just being superimposed over the user's field of view) to content creation, while also including occlusion (digital objects being naturally covered when any physical object is placed between them and the user), greatly increases the sense of realism and presence these digital objects have.

This also means there are now new interaction mechanics you can take advantage of when creating AR experiences for your business. For instance, if you were encouraging people to engage in fun treasure hunt activities to collect coins, you could hide them behind real-world objects.

Much of the AR technology we have experienced to date has been via applications installed on our mobile devices, but some companies have recently developed ways to deliver AR experiences via mobile web browsers without the need to install an app, thereby massively reducing friction for AR content consumption. For example, a company called 8th Wall, recently acquired by Niantic, has developed a way to deliver AR content via mobile web browsers, with the content downloaded in real-time and overlaid on the user's mobile phone camera view. This means you can click on an advert and get rich AR content on your mobile phone. And, in the future, when AR glasses become more widely used, this content will be displayed on top of your field of vision.

AR is already being used to promote TV shows and movies. Look at how *Game of Thrones* used Snapchat's Landmarker AR Lens to make a dragon land on top of the Flatiron building in New York in 2019, with the promotion launched to coincide with the premiere of season eight of the show[2].

What is currently holding AR back from taking over the world is the user experience, which is still far from optimal, because we still need hand-held devices like our mobile phones to view AR content. AR glasses will significantly improve the user experience – and, when the new wave of these is released, we will see mass adoption of AR both by consumers and businesses.

Brand affiliation with cryptocurrency

The way that cryptocurrency currently works, brands have the ability to print money from thin air, using their brand to provide the underlying value. Here's how this could work. Imagine a scenario where a luxury brand decides to create a cryptocurrency that carries its branding, allowing users to spend this digital currency to purchase products and services from that specific brand. If a well-known luxury brand announced that it was launching its cryptocurrency with a value of $0.05 per coin initially, how many people do you think would jump on that? Almost certainly enough for it to gain traction. As more people buy these branded coins, their value increases. How does this benefit both consumers and the brand? First, the brand could offer discounts or incentives to consumers who buy their products using their branded cryptocurrency. Second, the brand could create a product that is only available to consumers who pay with the branded cryptocurrency.

In addition to attracting consumers to the brand, this has another benefit for the business. Because the whole transaction is digital and internal to the brand, there are no fees to pay on these transactions. People will be paying the brand directly for their cryptocurrency, and, as its value rises, the brand receives more money with each of these purchases. This is one of the many opportunities that brands have right now to capitalise on their brand value in this digital world. It is huge – and, at the time of this writing, most of the markets are unregulated, which means it's possible to experiment and try some very interesting things.

Of course, the earlier you jump into these opportunities, the higher the risk if they don't work, but the higher the reward if they do. As long as you approach these opportunities with a strategy and do so in an informed way, it is a very exciting time for businesses to get involved.

A business' guide to starting out in the Metaverse

The first thing you need to know before you get into the Metaverse is who your target audience is. You also need to be clear about what you want to achieve by being in the Metaverse – are you trying to gather people who have the potential to be immediate customers, or do you just want to raise brand awareness? Many of the users on *Roblox*, for example, are children and teenagers. They might not use your products or services today – but, by having a presence in *Roblox*, you can introduce your brand to them to encourage them to come to you in the future.

If you are aiming for the first option – to attract customers immediately – the key is to make sure the experience you provide in the Metaverse contains as much information about your product as possible. You could also consider attaching to your physical offering a digital product or digital service that is available in the Metaverse.

If, however, you are aiming for the second option, it is essential to make any interactions with your brand as fun as possible. You want the young people using this platform to associate your brand name with your service, but also with those positive feelings of fun and enjoyment. If you're a global bank, for example, you could sponsor a Justin Bieber concert and have your logo displayed above the stage. When the young people who attended that concert are old enough to open a bank account, they will see that bank's brand, make a positive association and be much more likely to choose this bank than any of their competitors.

When is the right time to get into the Metaverse?

The quick answer to that question is 'yesterday!', although there is, of course, more to it than that. We would say that the right time to get into the Metaverse is as soon as possible, provided you approach it in the right way and do the right thing. Now, this is where it becomes a bit more of a difficult question to answer, because the right thing to do will vary depending on both what you have to sell and what your objective is.

It can help to answer the following questions to narrow down where your focus needs to be:

- Do you want to enter the Metaverse to make money?

- Do you want to enter the Metaverse to gain visibility, and you'll worry about making money in three years?

- Do you want to enter the Metaverse to capture a segment of consumers that you don't currently appeal to?

- Do you want to enter the Metaverse to quickly, collaboratively and iteratively test new products and services?

Step 1: Make sure your goal is crystal clear

There is nothing worse than a company entering the Metaverse thinking they are just going to make some noise only to be disappointed when they don't make any money. If you do not set out to make money, that won't be a surprise!

Step 2: Know what you are testing

If you decide to approach the Metaverse as a test, know what you are testing for. You need a strategy. If you are entering the Metaverse to capture 13–17-year-olds, dropping NFTs that cost $20,000 doesn't make sense, because the people in that *demographic don't have that kind of money. You have to think strategically about how to achieve your goals.*

Step 3: Know when is the right time

While we believe it is advisable for businesses to enter the Metaverse sooner rather than later, this will depend on your target demographic. For instance, if you are targeting 60-year-olds, you may need to wait a while, because they will not necessarily engage with this platform in the immediate future.

Step 4: Research the different players in the Metaverse

The Metaverse is made up of many different platforms, so you need to decide which will be most appropriate for your brand and which will help you achieve your goals most effectively. At the time of this writing, Roblox is one of the largest Metaverse platforms, with a user base of 50 million daily users, whereas Decentraland has around 18,000 daily users. This means you may want to wait and see which platforms emerge and present the strongest opportunities for your brand and business.

Step 5: Make your strategy consistent with brand messaging and mission

Whatever strategy you adopt in the Metaverse, it needs to be consistent with the strategy you are using in the physical world. Nike is a good example of how this can work. By creating Nike Land in the Metaverse, the brand is carving out its space. Similarly, in the real world, Nike is increasingly pulling its products from other retailers and only making them available through its own stores. Time will tell if this overall strategy pays off for the brand, but the consistency across mediums is certainly helping at present. Nike is putting out the message that it is there, but that you have to go to it.

Gucci has taken a different approach, they first started with what we describe as a 'granular strategy' in the Metaverse. Instead of immediately having their own 'kingdom', Gucci decided they first wanted its products to be available in any luxury focused corner of the Metaverse; and only recently they opened their own "Gucci Garden" on Roblox to showcase their products similarly to what Nike is doing with Nike Land.

If your brand message is all about having fun and enjoying yourself, it makes sense to connect it to music events or parties happening in the Metaverse. When you are considering various strategies for entering and scaling your business within the Metaverse, make sure you keep coming back to your existing brand mission and messaging.

We are still discovering the Metaverse

In Chapter 2, we discussed the concept of digital primitives and explained that many of the use cases for the Metaverse have yet to be uncovered. We're on the edge of that greyed-out map, and we have no idea what the rest of the landscape looks like yet, but the only way to find out is to start exploring.

As a business looking to enter this space, it is important to recognise that we still do not know which use case (or cases) will be most successful in the future. We are already seeing some use cases becoming more successful than others, such as art or digital goods being sold via NFTs, but we are still just scratching the surface of what is possible. I envision that, to really see the birth of disruptive use cases, we will probably have to wait another five (maybe ten) years.

That said, any company that starts investigating potential new use cases within the Metaverse now will have an advantage over businesses that wait and see, because the companies that embrace the Metaverse now will have the opportunity to explore and understand the technology in such a way that they can identify, create and design those transformational use cases. Look at Tinder, which has become synonymous with innovation in dating apps. Yet, before its launch in 2012, this use case was not something that anyone else envisaged – swiping left and right on strangers' pictures to get a match and maybe a date was a completely alien concept – and now it's how many people meet their partners.

We are still waiting to see what the emergence use cases in the Metaverse will be. Tinder was so innovative that it created a new genre of apps. We will see something similar happening within the Metaverse, where companies will create use cases that are so innovative and unique that they will become genres of their own. We have seen a similar pattern of innovation within the world of video games, where *Doom* created the first-person shooter, and *Ultima Online* created the massive multiplayer online role-playing game genre. The point is that all of these innovations started with companies that were bold enough to experiment with new technology, combining multiple experiences, designs and strategies.

By starting to experiment now, you are giving your company the best chance of becoming proficient in the use of the technology and thereby more opportunities to find those innovative and unique use cases that will emerge as we use the Metaverse more and more.

Endnotes

1. Sutcliffe C (2022) 'Nearly 7 million people have visited Nike's metaverse store', *The Drum*, 22 March, available at: https://www.thedrum.com/news/2022/03/22/nearly-7-million-people-have-visited-nike-s-metaverse-store
2. Spangler T (2019) '"Game of Thrones" Ice Dragon Lands on NYC's Flatiron Building in New Snapchat Lens (Watch)'. *Variety*, 12 April, available at: https://variety.com/2019/digital/news/game-of-thrones-snapchat-dragon-flatiron-building-1203186788/

CHAPTER 9
THE KEY ROLE OF ARTIFICIAL INTELLIGENCE AND DEEP LEARNING FOR THE METAVERSE

Deep learning, generative adversarial networks (GAN) and neural radiance fields (NeRF) fall under the umbrella of artificial intelligence (AI). There are multiple objectives we can achieve by harnessing this technology, particularly in the Metaverse, which present fantastic opportunities as well as potential risks. Let's look at what these might be, and what we can do to seize the opportunities and mitigate the risks.

Application #1: Content generation

It will not be long before the demand for content exceeds supply, which means both creators and companies will need ways to automate the generation of 3D content and assets. Technologies like GAN and NeRF can provide these kinds of generative AI services and do so in a way that produces incredibly realistic results.

How might this work in practice? Imagine creating 3D models using a vast database of 2D images that can be fully or partially ported into a model by AI. This will be particularly useful for companies operating online stores in the Metaverse, because they will be able to create a 3D model of a store using 2D images and AI technology. In fact, photogrammetry already exists and is used in everything from real estate and engineering to forensics and entertainment. However, it requires a large amount of visual data to produce an accurate 3D model.

With 3D GAN, a complete 3D model can be generated with just a couple of pictures, because the AI is able to analyse the images it is provided with and use this data to create replicas of what is within those images. NeRF, meanwhile, uses a process known as *inverse rendering*, where the AI is able to approximate how light behaves in the real world based on a handful of 2D images taken from different angles, enabling the AI agent to create a 3D scene that fills in the gaps from those images.

You can see how this technology could easily be applied in Metaverse platforms where we are creating landscapes, buildings and whole new worlds to explore.

Application #2: More realistic avatars

An avatar is the digital persona of a real user in a virtual world. AI can be used to create more realistic avatars, particularly in relation to their facial expressions and inverse kinematics, which relates to the avatar's movement. Inverse kinematics can be used to track the movements of a user and translate those movements in the real world into movements performed by the avatar in a virtual world. If you look at the Metaverse platforms we have at the time of this writing, the majority of the avatars you come across don't have legs (they are literally floating torsos and heads that are cut off at the waist – weird!), because this movement is especially difficult to translate into 3D worlds in a way that's believable.

With AI technology, there is the potential to make the avatars we have in virtual settings much more realistic, both in terms of how they move and how they express emotions.

Application #3: Autonomous AI driven NPCs

We already mentioned that AI could provide more realistic avatars for real humans, but the data collected for that purpose can theoretically also be used to train AI agents that for fully automated and photorealistic and virtual humans that look and behave much more like humans. There are advantages and risks to this, however.

The advantages are that we will be able to create much more believable non-playing characters (NPCs) in the Metaverse. These completely autonomous NPCs could act as guides within the Metaverse, helping new users learn how to navigate different platforms as well as provide ongoing support. Think of these NPCs like an avatar version of Siri or Alexa, who can act as a concierge in your virtual world.

Of course, the risk to this is that these avatars could be used to manipulate us, either by encouraging us to buy certain products or even change our political views, as I explained in Chapters 3 and 7. This is why it is very important for us to consider what biometric data can be measured by the organisations running Metaverse platforms in the first place and how any data that is measured is used.

While there is the potential for this data to be misused, I predict that the increased use of Metaverse platforms and immersive headsets (both augmented reality [AR] and virtual reality [VR]) will lead to immense data lakes of human behaviour that could potentially lead to the creation of better-quality automated robotics, and possibly even allow us to realise the dream of having human-like, fully autonomous robots within the next century.

When you think about it, this makes sense, because robots behave using vectors of movement in space. Take an automated arm in a car factory – this arm is using vectors of movement to understand where it has to solder on the different parts of the car. In order to input those vectors, you need precise data. However, to create believable movements in a human, you can create reinforced learning algorithms that can understand the position of a human body in different poses.

We are already seeing this happen in video games. Just look at how the game publisher Electronic Arts has used inverse kinematics of football players performing different actions to vastly improve the movements of the players in its *FIFA*-series football games. The company's AI then works to create much more believable avatars on the virtual pitch.

Application #4: Automatic detection of antisocial behaviour and harassment in the Metaverse

One of the wonderful things about the Metaverse will be its diversity. People will be able to express themselves however they want. Of course, there are some people in this world who don't believe in freedom of expression and who will disagree with other people's life choices, which could give rise to harassment and antisocial behaviour on Metaverse platforms.

It is already proving difficult to police the Metaverse, using the standard approaches taken by the likes of social media, such as checking images and written content for anything offensive. In Chapter 7, we explored harassment on Metaverse platforms and some of the steps that can be taken to prevent it, but the more users we have in the Metaverse, the harder and harder this will be to keep track of.

This is where AI can play a role, because it can be trained to look out for inappropriate behaviour between people's avatars in the Metaverse and do so much more effectively than a team of humans ever could. Of course, users will always be able to report inappropriate behaviour, but having an automated system to detect harassment and inappropriate behaviour in the Metaverse will be very useful and, I believe, very much needed.

Application #5: Faster and more accessible world creation in the Metaverse

As I write this, to create or edit worlds in any Metaverse platform, you need some development and coding skills, because you have to use the Unreal or Unity 3D engines for world building. In the future, however, I envision that the development and alteration of 3D worlds in the Metaverse will become increasingly seamless and simple to do. There are already platforms, such as Dreams for PlayStation, that are exploring a low-code and even a no-code interface for world building.

I believe this will become even simpler, to the point that users will be able to change what they can see in certain Metaverse environments through simple voice commands. Meta is already experimenting with this, and, in a demo[1], Mark Zuckerberg created a seaside scene, which was entirely AI generated, through a series of voice commands (it might not be the most realistic or inspiring beach scene at the moment, but it shows the potential of the technology).

This potential is actually mind-blowing when you stop to think about it. Picture yourself on the 'Holodeck' on the *Starship Enterprise*. You ask it to take you to a 1930s jazz club in Chicago, and all of a sudden you're immersed in a dark, moody bar. Smoke curls through the single spotlight that's focused on the piano on a small stage, with the pianist having a drag of his cigarette before sitting down to play. As the music floats through the room, you turn and look behind you to the bar. One bartender is making a drink for another customer, while the other is lazily cleaning the countertop while watching the performance. If you're a *Star Trek* fan, you no doubt have your own idea about which place and time you'd ask the Holodeck to recreate for you.

I believe, in the next 20 to 30 years, we can expect this level of sophistication in content programmatically generated based on user requests. Naturally, what experiences we can have and what situations and places we can recreate will depend on the data that is fed into the AI that sits behind and powers the platform. We've reached a critical mass with AI technology though; the snowball is rolling down the hill, and it's only going to gather pace.

Application #6: Moving digital assets seamlessly between Metaverse platforms

In Chapter 3, I explained how there is work being carried out to set consistent design standards across different Metaverse platforms, with the aim of allowing us to move, with our digital assets, seamlessly from one platform to another.

This is particularly important for any brands that sell digital assets, whether that's a designer handbag or a flashy sports car, because having the ability to take them into different worlds will matter to many Metaverse users. For example, if you sell a non-fungible token (NFT) for a Gucci handbag that can only be used in *Roblox* and not transferred with your avatar to *Decentraland*, it could make the NFT less appealing to consumers, and therefore less valuable.

Technology such as GAN and deep learning could be utilised to programmatically create 3D filters that will automatically generate a version of any digital asset you hold in the graphical style of multiple platforms. Your Gucci bag might look a bit different when you're carrying it in *Sandbox* as opposed to in *Roblox*, because it will be rendered in a way that is compatible with the style of that platform, but it will still be your Gucci bag because of the NFT it is attached to.

AI is becoming increasingly sophisticated

As you've read about some of these applications, you might feel as though they are years, if not decades, away from being realised, but AI is more sophisticated than you might imagine, and already incredibly accurate when it comes to creating photo-realistic images. Research published early in 2022 found that AI-synthesised faces are already almost indistinguishable from real faces and, perhaps more interestingly, are also considered to be more trustworthy than real faces[2].

This highlights where there are risks with the adoption of AI technology within the Metaverse, because if we can already generate not only photorealistic, but also more trustworthy, human faces, we have to consider how these could be used for nefarious purposes in the Metaverse, as I explained in Chapter 7 when we discussed data ethics within the Metaverse.

You can't talk about AI without talking about data, because, for AI technology to be effective, it needs as large a pool of data as possible. This data also has to be in the right format, so an AI that is able to write needs text-based data, an AI that creates 2D images needs 2D image data and so on. To create believable 3D worlds, these algorithms also need 3D content, which is increasingly easy to collect, now that the likes of the iPhone 12 Pro (and later iPhone Pro models) even come equipped with a light detection and ranging (LiDAR) scanner (3D scanner).

In fact, 6D.ai (which was acquired by Niantic Labs in 2020) set out to use this technology to create a persistent map of the world. Since being acquired by Niantic, the company's software development toolkit (SDK) has been injected into Niantic's SDK called Lightstrip. This is significant because Niantic is the organisation behind *Pokémon GO*, which means that every person playing *Pokémon GO* around the world is now contributing to creating this 3D map. This has also shifted Niantic's business model, because the company is now making its SDK available to anyone who wants to create a game or application that is map based.

The key when it comes to creating 3D worlds using AI is to ensure that all the specific assets within any 3D world are semantically segmented, which essentially means they should be correctly labelled – for example, as a bicycle, a tree, a child and so on. This is important because, once an image is segmented and injected into an AI algorithm, the algorithm can start to understand the common characteristics of assets with the same semantic label, and this allows it to start producing other assets that are similar to, or a variation of, that asset.

We also can't forget about GAN, which is an AI algorithm that is literally designed to fill in the blanks, as I explained earlier in this chapter. Nvidia has found a very interesting use for a GAN AI algorithm, namely, to produce super-resolution images in place of low-resolution images. Its GAN algorithm is essentially able to take images that have a resolution of 1080p and scale them up to, potentially, 4K resolution. To do this, the Nvidia algorithm is filling in the blanks using a hallucination based on what the AI agent computes should be there due to pixel resolution. This means, if you are playing a game on a lower-spec computer with 1080p resolution, you can upscale the images and run it in 4K resolution without using the graphics processing unit (GPU). Instead, you are using an AI agent that is running on the GPU, but that uses a fraction of the power to compute a 4K image. This technology of AI super resolution could be one of the key ingredients to achieving high-quality graphic renderings on lower-spec devices such as mobile phones or self-contained VR headsets such as the Meta Quest 2.

Holding the power of creation

The Metaverse has the potential to give us the opportunity to be almost God-like in the way we create and adjust our world. You could walk into an empty room in the Metaverse, say the words 'Let there be light!' and light will appear. Similarly, you could command trees, furniture and whatever else you want in your space, simply through the power of your voice and the AI algorithm that interprets and realises your commands.

These tools can make us into God-like creators, which can be wonderfully empowering. Of course, there are risks associated with this, as we have discussed, and if it is not regulated effectively, we face the slightly terrifying possibility of entering *The Matrix*, and a world that is so compelling and designed to trigger our dopamine receptors that it will always make us happy and trap us in a digital reality.

We want AI to enhance our lives, and the Metaverse has the potential to play a huge role in that, but we don't want to be bewitched by AI algorithms that can outsmart our capacity for discerning fiction from reality and genuine behaviour from deception. We are standing at the dawn of a new era, and we have the opportunity to ensure that technology is used to empower and enhance our lives, rather than to manipulate and control them.

Endnotes

1. Meta AI (2022) *Builder Bot Demo*, YouTube, available at: https://www.youtube.com/watch?v=62RJv514ijQ
2. Nightingale SJ and Farid H (2022) 'AI-synthesized faces are indistinguishable from real faces and more trustworthy', *PNAS*, 14 February, 119 (8) e2120481119, DOI: https://doi.org/10.1073/pnas.2120481119

CLOSING THOUGHTS

The Metaverse is an evolution of the Internet. It is a layer of infrastructure that will enhance the way the Internet interacts with the real world. It presents many opportunities as well as many challenges, and while some of us may be aware of the impact it can and will likely have on society, there are many people, including governments, who remain completely unaware of the direction we are travelling in.

As I write this, the Metaverse is still very much a work in progress; it is a vision, an ideal infrastructure that will enable us to be present with our friends and family anywhere in the world. It will allow us to live adventures in fantasy lands and to work remotely while feeling like we are present with our colleagues. Technology like digital currencies has the potential to enable people who are not living in countries with well-developed local economies to earn good money.

However, not all of this is possible yet. When we have these discussions about the Metaverse and what is possible, it is a little bit like the conversations people had back in the 1980s and 1990s about mobile phones. Back then, nobody could envision the impact that smartphones would have on society and the world just 10 years later. People had ideas about how mobile technology could be used. I remember people talking in the 1990s about making payments using mobile phones when most of us still carried cash around. Now, of course, we have the likes of Apple Pay and Google Pay on our phones, which has made that vision a reality. I sometimes leave the house without my wallet, because all I need now is my smartphone, or even just my smartwatch.

The COVID-19 pandemic has further accelerated the rate of technology adoption around the world. For example, before 2020, just a small percentage of purchases in South East Asia were made using credit cards or mobile payments. Owing to the pandemic, this figure has reversed, and now the vast majority of all transactions in South East Asia are made via digital transfers. This has forced businesses to adapt very quickly and undergo a digital transformation to enable them to take payments in this way.

This shift in approach to payments also implies that these businesses could now start delivering their products and services digitally. The pandemic hugely accelerated our adoption of technology, and, in all honesty, had it not been for the COVID-19, I don't think we would be talking about the Metaverse in these terms yet.

The various lockdowns in countries all over the world made people crave social connections, and, when meeting in person was not possible, the only way to do so was digitally. This drove adoption of platforms like *Roblox*, *Sandbox* and *Decentraland*, pushing their user levels to a critical mass. Similarly, the digital economy exploded, not only in terms of cryptocurrencies, but also the significant uptick in the adoption of digital payment methods. It was the perfect storm of global events and the rapid adoption of technologies that brought us to where we are today.

The consequences that this rapid digital transformation has for the world in which we live will depend very much on how we treat and use the new technology that we now have access to. We need to be responsible. We need to be ethical. We need to be inclusive. We need to think not only about minorities, but also about the digital divide.

Learning lessons from the 1990s

Those of us who are involved in the Metaverse already, whether as business technology providers, platforms, strategists, consultants or systems integrators, know that there is a great deal that needs to happen to ensure that this new technology supports and enhances our society rather than destroys it. We are the people who need to help governments understand that the Metaverse is not just a fad, but simply the inevitable evolution of the Internet.

If we want to have a Metaverse that represents an idyllic utopia, rather than a dystopian environment, there are some practical problems we need to solve and risks we need to mitigate. These include challenges around diversity and inclusion, accessibility, safety, fake news, freedom of speech and mass manipulation. All of the challenges I have outlined in this book need to be tackled before we can create the utopian Metaverse we imagine in our heads. If we don't deal with these issues now while we are designing the foundation of the

Metaverse, they will surely remain endemic and spread across all Metaverse platforms, creating a dystopic scenario in which mass manipulation, harassment, exploitation of user data and inequality will be common.

We cannot forget the lessons from the 1990s, when the Internet was seen as a beautiful and wonderful force that could unite the world and provide knowledge to everybody. We can see how that story ended very differently. There are governments exploiting mass media communications to influence the political opinions of entire populations. There are criminals who hack and carry out scams. There are online trolls who hide behind their anonymity to spread hate and malevolence in the world.

In fact, of all the challenges I have highlighted in this book, anonymity online and in the Metaverse is one of the most pressing ones we need to address. We have to find a way to create accountability across Metaverse platforms and to enforce the law to ensure we create a space where people feel safe to be themselves and interact with others. There is too much anonymity on the Internet at present, and this prevents law enforcement agencies from taking any real action against people who harass and threaten others online. I predict that having a way to uncover people's digital identities and to monitor their behaviour online and across Metaverse platforms will be a huge topic of discussion in the years following the publication of this book.

Another discussion running alongside this will be around the laws and regulations that govern Metaverse platforms and how we can enforce them. This is an essential aspect of the ethical debate, because the Metaverse will be the bridge that connects the physical and virtual worlds. Therefore, it's important to find a way to introduce the same level of laws and governance to the Metaverse as we have protecting the very fibre of our society in the real world.

The key is to start a dialogue around these issues, not only among those of us already using and starting to understand the Metaverse, but also with those in government who have no idea of the scale of impact this will have on our lives. I am talking about building the Metaverse in a responsible way, and it is up to those already operating within the Metaverse to take the lead.

There are already non-profit organisations and consortia working to develop standards to enable interoperability across Metaverse platforms, but what we are lacking are similar organisations working in the same way in relation to the ethics of the Metaverse.

Creating guidelines and working closely with local governments, as well as supra-national organisations like the EU and UN, to enforce those guidelines will be a key steppingstone to ensure that the Metaverse is a safe place for everyone. We do not want to have one large corporation dictating the rules in the Metaverse, simply because it has the best standards and the best technology. How we live and interact across these platforms should be determined by those of us building the Metaverse as well as those who will be using it (which, if you haven't realised by now, is all of us!).

It is already clear what scenarios we don't want to see. We don't want brands to be able to access our biometric data – otherwise, for example, they will be able to see our heart skip a beat when we see a digital version of a sports car we like passing by one of the Metaverse platforms and therefore target us with advertising for that specific car in the real world with the purchase only "one click away" or, possibly worse, for a loan so we can afford one! Or for insurance companies to use our biometric data to predict the likelihood of us suffering from a disease in future and therefore inflate our premium quotes because they already know we are going to get sick, even if we don't! We can avoid these situations from happening if we put the right regulations in place now.

Are you a trailblazer or a laggard?

When it comes to new technology, there will always be trailblazers and laggards, and there are some considerations to bear in mind whether you belong to either group (if you have a PlayStation Vita, GoPro Karma or an Oculus Rift in a cupboard somewhere, chances are you're either a trailblazer or a technology reviewer). For example, trailblazers could start a project in the Metaverse only to fail spectacularly, because either they don't have enough knowledge or they haven't chosen the right time for their business and brand to enter the Metaverse. The flip side to this is that trailblazers will acquire the know-how, experience and capability to work within this new environment more quickly.

This highlights the major downside for the laggards, namely that, by the time they enter the Metaverse, they could be so far behind their competitors who got in earlier that they are simply unable to compete. You don't want your business to be like Kodak, who only started exploring digital cameras after they had been available for nearly a decade. This means that, even if you don't fall in the trailblazer group, you at least need to have a strategy for how your business will enter the Metaverse, and to lay out clear pros and cons for deciding the timing for your entry.

Whichever group you fall in (or maybe you land somewhere in between), it isn't expensive to enter the Metaverse; in fact, businesses can do so for just a few ten thousand dollars. If you are going to get involved, you need a strategy not only around when and how you will enter, but also around how you will build your community and offer services around it.

Given how little it costs to start experimenting in the Metaverse, it is worth getting involved sooner rather than later so that you can start finding your audience and working out how best to engage with them. Also remember the high footfall we are already seeing on certain Metaverse platforms, such as *Roblox*, which has over 50 million unique visitors a day at the time of this writing (2022). There is huge potential here for people to see and engage with your brand and products.

As we saw in Chapter 8, there are a multitude of ways in which you can enter this market, and, if you do this well, you can significantly grow your business. However, regardless of whether your business is interested in cryptocurrencies, NFTs or virtual land, you need to enter into it from an informed position. You've made a good start by reading this book, but don't stop here! Read more on your subreddit; watch YouTube tutorials;

ask the experts; join communities; do your own research. This is how you will continue to explore your Metaverse map and open up new frontiers for your business.

Investing in the Metaverse is like investing in any other asset – in that you need to carry out thorough research, or you risk losing a lot of money. However, if you take a strategic approach and arm yourself with good information before you jump in, you stand to gain a great deal.

We can't see the edge of the map

The Metaverse is an opportunity with endless possibilities. The map doesn't suddenly end; it will just keep revealing itself to us as we continue to explore it. It is not only by creating virtual worlds that we can have an impact; this can and will change the real world too. We are on the verge of being able to connect both our physical and digital realities in such a way that we can move seamlessly between them, and that is incredibly exciting.

As a business, this presents opportunities to open up entirely new markets, even create new business models, and to reach new audiences. However, you have to ensure you are entering the Metaverse with a strategy that has a clearly defined business goal (your North Star) that you can always follow. Ask the right questions to make sure you create your strategy from a place of knowledge rather than from a place of assumptions.

The new use cases and ideas for the Metaverse will come from strategic thinking and by prioritising your ideas according to impact and feasibility. The most impactful and the most feasible ideas you come up with are your no-brainers – do these immediately! Once you have a list of other ideas, prioritised according to those two criteria, you have a solid base for a strategy that you can break into actionable steps. This is your map into the Metaverse, so make sure you have well-defined key performance indicators (KPIs) to allow you to measure your success and progress along the way.

My advice to you, whatever business you operate, is to be bold and brave. We are the pioneers of this new world, and there is a great deal still to discover. No doubt, there will be some surprises along the way – hidden treasures in some places on this as-yet unknown map, and dragons in others – but fortune favours the brave. If you would like to contact me to have a conversation about the Metaverse, how best to enter it or any of the issues I've raised in this book, you can connect with me by scanning the following QR code. I wish you all the best as you embark on your journey and start turning that greyed-out map into a living, breathing world full of possibilities.

Let's connect!

ABOUT THE AUTHOR

Nicola 'Nick' Rosa

Creative technologist, experience designer and digital transformation specialist who previously worked for Spotify, IBM IX, Yahoo!, and Atari – Nicola 'Nick' Rosa is now part of Accenture Metaverse Continuum Business Group, where he leads the Metaverse and Extended Reality Strategy in Europe.

Nick is also the chairman of the British Interactive Media Association (BIMA) Council for Immersive Technologies, where he helps define best practices, use cases and the advantages of using XR for UK digital agencies. He is a director at the Academy of International Extended Reality (AIXR), a non-profit entity producing the annual VR Awards. Nick is also the co-host and co-author of a monthly podcast produced by AIXR called 'Field of View', where he interviews entrepreneurs, researchers and pioneers in the fields of the Metaverse and Extended Reality.

INDEX

INDEX